MANUAL

OF

FIELD ENGINEERING.

1911.

GENERAL STAFF, WAR OFFICE.

Fredonia Books
Amsterdam, The Netherlands

Manual of Field Engineering 1911

by
General Staff, War Office

ISBN: 1-58963-763-1

Reprinted from the 1911 edition

Fredonia Books
Amsterdam, the Netherlands
http://www.fredoniabooks.com

CONTENTS.

CHAPTER IV.

IMPROVEMENT OF THE FIELD OF FIRE AND UTILIZATION OF EXISTING COVER.

CHAPTER V.

EARTHWORKS.

CHAPTER VI.

OBSTACLES.

CHAPTER VII.

DEFENCE OF LOCALITIES AND POSITIONS.

CHAPTER VIII.

FIELD LEVEL AND FIELD GEOMETRY.

CHAPTER IX.

CAMPING ARRANGEMENTS.

CHAPTER X.

KNOTTING AND LASHINGS.

CHAPTER XI.

BLOCKS, TACKLES AND USE OF SPARS.

CHAPTER XII.

COMMUNICATIONS.

CHAPTER XIII.

BRIDGES AND THE PASSAGE OF WATER.

CHAPTER XIV.

HASTY DEMOLITIONS WITH EXPLOSIVES

CHAPTER XV.

HASTY DEMOLITION OF RAILWAYS AND TELEGRAPHS WITHOUT EXPLOSIVES.

APPENDICES.

DEFINITIONS
AND
TECHNICAL TERMS.

Abatis.—An obstacle formed of trees or branches of trees picketed to the ground, with their points towards the enemy.

Angle of Descent.—The angle which a tangent to the trajectory makes with the line of sight at the point where they intersect.

Banquette.—The place upon which men stand to fire over a parapet.

Batter.—The slope of the face of any stone or masonry structure which is not vertical.

Baulk.—A prepared roadbearer in a military bridge.

Bay.—The distance bridged by one set of *baulks* or roadbearers, i.e., the distance between any two *transoms* or transverse supports measured from centre to centre.

Berm.—A small space left between the parapet and excavations of a work.

Bight.—A loop formed on a rope, the two parts of which lie alongside one another.

Bivouac.—An encampment without tents or huts.

Bomb-proof.—A shelter, proof against the penetration of shells.

Butt.—The larger end of a round spar.

Calibre.—The diameter of the bore of a gun in inches, measured across the lands.

Caponier.—A small chamber formed in the ditch of a work and projecting from the escarp to give fire along the bottom of the ditch.

Casemate.—A shell-proof chamber constructed for the accommodation of the garrison of a work or position.

Chess.—A prepared plank forming a portion of the flooring of a bridge.

Choke.—The mouth of a sandbag when closed and secured.

Command.—The vertical height of the crest of a work above the natural surface of the ground, or above the crest of another work.

Cordage.—Hemp or fibre ropes.

Counterscarp.—The side of the ditch of a work furthest from the parapet.

Cover.—Concealment from view or protection from fire, or a combination of both.

Crest.—The intersection of the interior and superior slopes of a parapet.

Crib-pier—A support for a bridge formed of layers of baulks of wood laid alternately at right angles to each other.

Dead ground —Ground which cannot be covered by fire.

Dead load.—A load which is applied to a structure continuously and which remains steady.

Defilade.—The adjustment of the levels of the crest and interior portions of a work with a view to obtain cover for the defenders or to screen them from view

Derrick.—A single or composite spar used for lifting or moving weights.

Detonator.—A small charge of fulminating explosive used to apply the pressure and temperature necessary to detonate a larger charge of explosive.

Direct laying.—The method of laying a gun by looking at the target over or through the sights.

Ditch.—An excavation which is not intended for occupation. It serves as a source of supply of earth for the parapet, an obstacle to assault, or a means of drainage.

Embrasure.—A channel through the parapet of a work through which a gun is fired.

Enfilade fire.—Fire, the line of which sweeps a target from a flank.

Epaulment.—A small parapet to give cover to a gun and detachment in action.

Escarp.—The side of a ditch nearest the parapet.

Exterior slope --The outside slope of a parapet extending downwards from the superior slope.

Fascine.—A long bundle of brushwood, tied up tightly, used for roadmaking, &c.

Field of Fire.—Any area of ground exposed to the fire of a given body of troops or group of guns.

Fire.—A military projectile during its flight.

Flèche or Redan.—A work consisting of two faces, forming a salient angle towards the enemy.

Foreground.—That portion of a field of fire lying nearest the origin of the fire.

Frontal fire.—Fire, the line of which is perpendicular to the front of the target.

Fougasse.—A small mine filled with stones which are projected towards the enemy on the mine being fired.

Gabion.—An open cylinder of brushwood, sheet iron, &c., used for revetting.

Glacis.—The ground round a work within close rifle range. It is sometimes formed artificially.

Gorge.—The face of a work least prepared to receive frontal fire.

Grazing fire.—Fire which is parallel, or nearly so, to the surface of the ground.

Gradient.—A slope represented by a fraction: *e.g.*, $\frac{1}{30}$ represents a rise or fall of one unit measured vertically for every 30 units measured horizontally.

Guy.—A rope fastened to the tip of a spar or frame to support, raise, or lower it.

Gyn.—A tripod with tackle used for raising weights.

Headcover.—Cover against frontal or oblique fire for the heads of men when firing.

Headers.—The term applied to sods, sandbags, &c., placed so that their longest side is at right angles to the face of the structure in which they are being used.

Helve.—The handle of an axe or pick-axe.

High angle fire.—Fire from all guns and howitzers at angles of elevation exceeding 25°.

Indirect laying.—The method of laying a gun when direction is obtained by an aiming point or aiming posts and elevation is adjusted by sight clinometer.

Interior slope.—The inner slope of a parapet extending from the crest to the banquette.

Keep or Réduit.—A separate enclosure within another work to enable the defenders to bring fire to bear upon a field of fire lying within the outer line of defence.

Lead.—Any line used to convey an electric current.

Ledger.—The lowest horizontal timber connecting the legs of a trestle or frame.

Live load.—A load which is suddenly applied to a structure or part of a structure with slight impact, producing strains in excess of those due to its weight when at rest.

Lunette.—A work consisting of four faces, the two centre ones forming an obtuse salient, the two side ones affording fire to the flanks.

Lunette, blunted.—A work consisting of five faces (otherwise similar to a lunette).

Machicoulis gallery.—A balcony with a bullet-proof floor and parapet: loopholed in the floor to afford fire in a downward direction.

Oblique fire.—Fire, the line of which is inclined to the front of the target.

Overhead cover.—Cover against projectiles whose steep angle of descent would otherwise enable them to strike the target.

Palisades.—Palings intended to resist the passage of troops.

Panjies.—Bamboo spikes, from 9 ins. to 1 ft. long, sharpened and charred to hardness. Driven about half their length into the ground and concealed in long grass, to form an obstacle.

Parados.—A traverse to give cover from reverse fire.

Pickets.—Posts sharpened at one end for driving into the ground by hand-power.

Pier.—A means of support intermediate to the abutments or shore-ends of a bridge.

Piles.—Posts sharpened at one end for driving into the ground by a force in excess of hand-power.

Plunging fire.—Fire when the target is much below the gun, so that the projectile strikes the ground at a steep angle of descent.

Primer.—A small piece of dry guncotton used to detonate wet gun-cotton.

Profile.—The outline of the section of a parapet at right angles to the crest.

Puddle.—Clay freed from stones and dirt, and worked up with water to form a plastic and waterproof lining to earthen reservoirs for water.

Racklashing.—A prepared lashing by which ribands are secured to the outside roadbearers of a bridge.

Rectifier.—A boxwood implement for enlarging perforations in gun-cotton primers so as to take the shanks of detonators.

Redan, blunted.—A work consisting of three faces, the centre one firing to the front, the others to the flanks.

Redoubt.—A field work entirely enclosed by a defensible parapet, which gives rifle fire all round. It may be of any command.

Relief.—The length of time that men have to work before being relieved, or a party of men who work, or who are on duty, for a given length of time.

Retrenchment.—A work or works arranged so as to form a second, but not necessarily separate, line of defence, and also usually to reduce the area to be covered with fire.

Revetment.—Any material formed into a retaining wall to support earth at a steeper slope than that at which it would naturally stand.

Reverse fire.—Fire which is directed against the rear of a target.

Riband.—A beam or spar fastened down on each side of a roadway to keep the chesses in place.

Roadbearer.—One of the longitudinal supports of the roadway of a bridge.

Rope.—Steel or iron wire rope.

Running end.—The free end of a rope; the rest of the rope is called the standing part.

Sap.—A trench formed by men working from the bottom of the trench and constantly extending the end towards the enemy.

Sangar.—A dry built stone wall to give protection against rifle fire.

Searching power.—The power of a projectile to reach an objective behind cover. It varies with the description of weapon, the range and the angle of descent of the projectile.

Sheers.—Two spars lashed together at the tip and raised to rest on their butts, which are separated. They are used to lift and move weights in one plane.

Span.—The horizontal distance between the centres of any two supports of a bridge. The length of a bridge from shore to shore is called the total span.

Spit-lock.—To mark out a line on the ground with the point of a pick.

Splinter-proof.—A shelter, proof against splinters of shell.

Storm-proof work.—A work is considered storm-proof when its design is such that, no matter how great the determination of the assailants may be, they can be destroyed as fast as they can advance to the attack.

Stretchers.—The term applied to sods, sandbags, &c., placed so that their longest side is parallel to the face of the structure in which they are being used.

Superior slope.—The top of a parapet immediately forward from the crest.

Tackle.—Any system of blocks and ropes by which power is gained at the expense of time (*i.e.*, more power—less speed).

Tambour.—A bullet-proof projection, constructed so as to flank the walls of a building.

Tamp.—To enclose a charge of explosive with earth or other material so as to confine the gases at the beginning of an explosion, and thus develop their forces more fully.

Task.—The amount of work to be executed by a man during a relief.

Tip.—The smaller end of a round spar.

Terreplein.—The surface of the ground inside a work.

Trace.—The outline of a work in plan.

Transom.—The transverse beam or support on which the baulks or roadbearers rest.

Trench.—An excavation which is for use either as a means of concealment or protection or both.

Traverse.—A bank of earth erected to give cover against enfilade fire, and to localize the bursts of shells.

Wattle.—Continuous brushwood hurdle work.

CHAPTER I.—FIELD FORTIFICATION.

(*See also* F.S. REGS., PART I, CH. VII.)

1. GENERAL INSTRUCTIONS.

1. By Field Fortification is implied all those measures which may be taken for the defence of positions intended to be only temporarily held. Works of this kind are executed either in face of the enemy, or in immediate anticipation of his approach.

2. Field Fortification presupposes a defensive attitude, and, though recourse to it may under certain circumstances be desirable, IT MUST ALWAYS BE REGARDED AS A MEANS TO AN END, AND NOT AN END IN ITSELF.

3. The principal aim of field fortification is to enable the soldier to use his weapons with the greatest effect, the second to protect him against the adversary's fire. By thus reducing losses and increasing the powers of resistance in any part of the theatre of operations or field of battle, more troops are available to swell the force destined for decisive action there or elsewhere.

4. The extent to which field fortification is employed will depend on whether a commander acts on the offensive from the commencement of an action, or whether he decides to await attack in the first instance.

5. If the offensive be assumed, field fortification will find only a limited application, for the provision of cover must never be allowed to stop the advance, and entrenchments will not be commenced without an order from an officer. During the process of establishing a superiority of fire, successive fire positions will be occupied by the firing line. As a rule, those affording natural cover will be chosen, but if none exist, and the intensity of the hostile fire precludes any immediate advance, it may be expedient for the firing line to entrench itself. (*See also* F.S. Regs., Sec. 105.)

If any diminution in the volume of fire is thereby entailed, infantry should only entrench when further progress has become impossible, and an energetic advance must be resumed at the earliest moment. Artillery should be entrenched whenever possible.

All important tactical points should, when captured, be at once put in a state of defence, so that attempts on the part of the enemy to recapture them may be defeated, and they may serve as supporting points to the attack. Local reserves will often find opportunities for strengthening localities or fire positions, which have been previously gained by the firing line. Detachments of engineer field companies may be attached to them to assist in such work.

6. On the defensive, the amount of work to be undertaken will depend on the object, the ground, the time, and the numbers of men and tools available. Other considerations which should guide a commander in the organization of a defensive position are dealt with in Sec. 56.

7. The following points should be borne in mind when examining a locality which it is desired to strengthen :—

(a) The strong and weak points of the position to be defended should be carefully studied, and the site for entrenchments chosen with due regard to tactical requirements and economy in men.

(b) The enemy in attacking should be exposed to the fire of the defenders, more especially for the last 300 or 400 yards. To ensure this, the foreground may require clearing.

(c) The enemy should be deceived as to the strength and dispositions of the defending troops, and the character of their works.

(d) The defenders should be screened from the enemy's view, and sheltered from his fire by natural or artificial cover, so arranged as to permit the maximum development of their own rifle fire.

(e) The free movement of the attacking troops should be hampered by obstacles to detain them under fire and to break their order of attack.

(f) The free movement of the defenders should be facilitated by improving communications within their position, and clearing the way for counter attack.

8. In order that field works may be designed to the best advantage, the effect of rifle and gun fire at various ranges should be fully realized.

2. RIFLE FIRE.

1. Modern military rifles are sighted to about 2,800 yards. Their maximum range may be taken as about 3,700 yards. The slope of descent of the bullet varies from about $\frac{1}{115}$ at 600 yards and $\frac{1}{70}$ at 1,100 yards to $\frac{1}{175}$ at 2,200 yards.

2. The heights over which an average man can fire on level ground as adopted in various armies, are :

	France.	Germany.	Russia.	Great Britain.
Lying down	—	11·8in.	—	1ft.
Kneeling	3ft. 3·3in.	2ft. 11·4in.	2ft. 10·8in.	3ft.
Standing	4ft. 7·1in.	4ft. 7·1in.	4ft. 8·0in.	4ft. 6in.

A higher parapet can be used when firing uphill than downhill.

8. The following table gives the maximum penetration of the pointed bullet in various materials.

In order to obtain proof cover, a percentage must be added to these numbers, *e.g.*, earth parapets should not be less than $3\frac{1}{2}$ feet thick. If the soil is free from stones, a thickness of 4 feet is desirable.

Material.	Maximum penetration.	Remarks
Steel plate, best hard	$\frac{7}{16}$ in.	At 30 yards normal to plate; $\frac{7}{16}$ ins is proof at not less than 600 yards, unless the plate is set at a slope of $\frac{2}{3}$ when $\frac{7}{16}$ in is proof at 250 yards.
ditto ordinary mild or wrought iron ...	$\frac{3}{4}$ in.	
Shingle	6 ins.	Not larger than 1 in. ring gauge.
Coal, hard	9 ins.	
Brickwork, cement mortar	9 ins.	150 rounds concentrated on one spot will breach a 9 inch brick wall at 200 yards ... ...
ditto , lime mortar ..	14 ins.	
Chalk	15 ins.	
Sand, confined between boards, or in sandbags..	18 ins.	Very high velocity bullets have less penetration in sand at short than at medium ranges.
Sand, loose	30 ins.	
Hard wood, *e.g.*, oak, with grain	38 ins.	
Earth, free from stones (unrammed)	40 ins.	Ramming earth reduces its resisting power.
Soft wood, *e.g*, fir, with grain	58 ins.	Penetration of brickwork and timber is less at short than at medium ranges.
Clay...	60 ins.	Varies greatly. This is maximum for greasy clay.
Dry turf or peat	80 ins.	

3. ARTILLERY FIRE.

1. *Field Guns.*—Both shrapnel shell and high explosive shell are fired by the field artillery of most foreign nations.

Shrapnel with time fuzes can be used up to a range of about 6,000 yards. With percussion fuzes shrapnel can be used effectively against troops behind 14 inch brick or 2 feet thick mud walls as they penetrate before bursting.

High explosive shell are intended for use chiefly against troops under cover or against shielded guns.

The angle of descent of the projectile varies from $\frac{1}{20}$ at 1,500 yards to $\frac{1}{4}$ at 4,000 yards.

2. *Field Howitzers* fire both shrapnel shell and high explosive shell. The chief difference between their fire and that of field guns is that the shell is heavier (varying from 30 to 45 lbs.), contains a larger

bursting charge, and has a steeper angle of descent. Howitzer fire therefore possesses greater searching power than that of field guns. They can fire shrapnel shell up to a range of about 6,000 yards, while their extreme range is about 7,000 yards. The angle of descent of the projectile may be as steep as ⅓.

The amount of cover necessary to keep out the shell of a howitzer is described in Mil. Eng., Pt. II, Plate X, Fig. 3. The effect of the burst, though powerful, is very local, and 9 to 12 ins. of earth or 3 to 4 ins. of shingle supported by some suitable material suffices against splinters from the shell.

3. *Heavy Guns and Heavy Howitzers* fire both shrapnel shell and high explosive shell of still greater weight at ranges up to 10,000 yards. These pieces are therefore specially useful for long range enfilading fire.

4. The aim of field artillery in the attack of a position is to assist the advance of its own infantry by bursting its shell in such a position that the defenders will either be forced to keep under cover or be struck. No attempt is made to breach the parapets of the defence.

An occasional shell may strike and penetrate the parapet, but in the case of shrapnel the damage to the parapet will be trifling, while in the case of a shell filled with high explosive, the effect will be no worse on a thin parapet than on a thick one. It is, therefore, useless to spend time and labour on making a thick parapet simply to keep out shell.

Plate 1 gives some idea of the effect of bursting shells.

5. The following is a table of ranges and the terms applied to them.

Terms applied to Ranges.	Rifle.	Field Artillery.	Heavy F. Artillery.
	Yards.	Yards.	Yards.
Distant	2,800 to 2,000	6,500 to 5,000	10,000 to 6,500
Long	2,000 to 1,400	5,000 to 4,000	6,500 to 5,000
Effective	1,400 to 600	4,000 to 2,500	5,000 to 2,500
Close	600 and under.	2,500 and under	2,500 and under

The extreme range of field artillery using percussion shell may be taken as 9,000 yards, and of heavy artillery as 10,000 yards.

The width of the area of ground struck by the bullets of an effective shrapnel is about 25 yards.

The length of the forward spread of the bullets of shrapnel burst at effective range is about 200 yards.

The radius of the explosion of a high explosive shell is about 25 yards.

CHAPTER II.—TOOLS, MATERIALS AND THEIR EMPLOYMENT.

(*See also* Military Engineering, Pt. I, Secs. 4, 5 and 6.)

For summary of tools and explosives carried on the person and in 1st line transport, see table, Appendix I.

For table of time, men and tools required for different natures of work, see Appendix II.

4. ENTRENCHING TOOLS.

(*See Appendix III, Table* 1.)

1. The service entrenching tools consist of *pick-axes, shovels, spades, crowbars* and the *entrenching implement.*

2. The shovel is used right or left handed ; and the thigh should be employed to assist in thrusting it under earth which has been loosened.

In throwing earth from the shovel there should be no jerk, the hand nearest the blade must be allowed to slide freely up the handle, otherwise the earth will scatter.

3. For safety in trench work the pick must be used working front and rear, and never sideways. Before striking the pick into the ground it should be raised well above the head with both hands. In bringing it down, the helve should slide through the hand nearest to the head, and full advantage taken of the weight of the pick-head.

4. The head of the entrenching implement has at one end a pointed shovel-shaped blade while the other is similar to the chisel end of a pick-axe. The head slips on to a helve 16½ inches long which is ferruled at one end. The whole tool weighs about 1lb. 18½oz.

In using the entrenching implement men should be taught to work lying down and to commence at the rear of the selected position. Hard soil is more easily broken up by this method and a hollow for the disengaged arm is gradually provided, which helps to keep the digger under cover.

5. CUTTING TOOLS.

(*See Appendix III, Table* 2.)

1. The service cutting tools are the *felling axe, hand axe, cross-cut saw, hand saw, folding saw,* and *billhook.* The importance of keeping all cutting tools sharp and firmly handled cannot be exaggerated.

6. EARTH.

1. Of all the materials, which are most generally available for the construction of field defences, earth is the most valuable as well as the most common. It is usually procured from trenches or ditches dug as near as possible to the place where it is to be used.

The steepest slope at which thrown-up earth will stand is about 45° or ⅓. To make it stand at a steeper slope it must be revetted.

7. SODS.

1. Sods are used for revetments and also to form walls in special cases. They should, if possible, be cut with a sharpened spade or sod-cutter from meadows growing thick grass and should be about 18 inches long, 9 inches broad and 4½ inches thick.

They must not be cut where the resulting scar will betray the existence of the works. Such a scar may sometimes be utilized to represent a dummy trench.

2. Sod revetment is built at a slope of ¾ (Pl. 2, Fig 8). The sods must be laid in alternate layers of headers and stretchers, grass downwards, breaking joint, *and at right angles to the slope*, with two layers of sods in each stretcher course. The top layer should be laid with grass upwards, and all headers. They should be bedded and backed by fine earth well rammed. For superior work a picket should be driven through each sod. Cleft fir pickets are better than round, which split the sods.

8. STONES.

1. Stones may be used to build rough walls in places where digging is difficult or impossible, or other material is lacking. Such walls should be about 2 feet thick at the top, with a batter not steeper than ¼ when dry-built. They are undesirable where artillery fire is expected, for the stones are apt to be thrown about when struck. Only the largest stones which can be handled should be used for such walls.

9. TIMBER.

(See Appendix III, Table 12.)

1. Timber is used in the construction of bridges, huts, splinter-proofs, stockades, abatis, &c.

2. The felling axe in the hands of an experienced workman is the best tool for felling timber. The hand axe is only suitable for felling trees up to 15 inches in diameter.

To fell a tree it should be strained in the required direction of fall by a rope. It is then cut into as far as the centre of that side, and finished off on the opposite side by a cut about 4 inches higher up.

3. If a saw is used for felling timber, the cut must be wedged open to prevent the saw being jammed.

4. Timber revetment is made by driving strong stakes into the ground, and placing planks or logs between them and the parapet. The slope will depend on the strength of the timber, ¼ will generally be safe. The stakes should be anchored back by holdfasts attached to them at about ⅔ their height out of the ground.

Planks should never be used for revetments where exposed to the fire of high explosive shells unless they are secured so firmly that they will break rather than come away whole.

10. BRUSHWOOD.

1. Brushwood is used for roadmaking, hutting and revetting. Willow, birch, ash, Spanish chestnut and hazel are the most suitable kinds, and work best if cut when the leaf is off.

As a rough rule it may be taken that 1,000 square yards of brushwood, up to 2 in. diameter, make up three G.S. wagon loads.

If brushwood has to be carried any distance it should be tied into bundles, weighing about 50 lbs. If nothing else is available, these may be bound with pliable rods called "withes," which should be well twisted before use. (Pl. 3, Figs. 1 and 2.)

2. To make a brushwood revetment, stakes are driven into the ground, from 1 to 2 feet apart, along the line of the intended parapet, and anchored back at a slope of about ¾, as described above. As the parapet rises, loose brushwood (or ferns, reeds, straw, &c.) is filled in between the stakes and the parapet. (Pl. 2, Fig. 2.) Stretching of the fastenings due to the pressure of the earth will make the slope somewhat steeper.

All anchorages for revetments must be prepared before, and not after, the completion of a revetment.

11. FASCINES.

1. A fascine is a long faggot tightly packed and bound, used for drains from trenches, foundations of roads in marshy sites, and occasionally for revetting steps, etc. The usual dimensions are 18 feet long and 9 inches in diameter. It is made in a cradle of trestles placed at a uniform level as shown on Pl. 3, Figs. 3 and 5.

Fascines make a poor revetment by themselves, and their use as a revetment is generally confined to steps. They should be well picketed down.

12. GABIONS.

1. Gabions are cylinders open at both ends, made of almost any material capable of being bent or woven into a cylindrical form, such as brushwood, canvas, expanded metal, wire netting, &c. Various patterns are shown on Pl. 4 from which the general method of construction can be obtained. They form a very stable revetment; but will generally be available in fortress warfare only. They can be used at any slope that may be desired by digging a sloping trench in which to place them. (Pl. 2, Fig. 5.)

2. Willesden paper band gabions are an article of store. Each gabion consists of 10 bands, 3 inches wide, fastened at the ends by two copper clips. (Pl. 4, Figs. 6 and 7.)

3. Willesden paper may shortly be replaced by "expanded metal." Gabions made of this material require no pickets (Pl. 4, Fig. 5). The material is also suitable for hurdles.

13. HURDLES.

1. Hurdles, unless for a special object, are usually made 6 feet long and 2 feet 9 inches high in the web. (Pl. 5.)

To make a hurdle a line 6 feet long is marked on the ground, and divided into nine equal parts, and a picket (about 8 feet 6 inches long and from 1 inch to 2 inches in diameter) driven in at each division, the two outside ones being somewhat stouter and longer.

The web is then constructed by a process called *randing* which consists in working with single rods commencing from the centre (Fig. 1). Each rod is taken alternate sides of the pickets, twisted round the end pickets, and woven back towards the centre. A fresh rod must overlap by several pickets the one which it supplants.

Pairing rods (*aa* Pl. 5, Fig. 2) are used in the centre and at both ends of the web, which is usually sewn top and bottom in three places.

2. A rougher type of hurdle is shown in Pl. 5, Fig. 3. This is made much more quickly than the ordinary type and is equally efficient for most purposes.

3. Hurdles are one of the most useful forms of revetment, either in the form of ready made hurdles, or continuous hurdle revetment constructed simultaneously with the parapet. In either case they should at first be placed at a slope of $\frac{3}{4}$, and frequently anchored into the parapet. Stretching of fastenings, due to the weight of earth in the parapet, will bring the hurdles to a slope of $\frac{4}{4}$, as shown in Pl. 2, Fig. 1. In continuous hurdlework the web is formed by *randing*, each pair of men having 10 feet or 12 feet of revetment as their task, and working in their rods with the men on either side. Hurdles are also used in the construction of huts and for temporary roadways.

14. Sandbags.

1. The service pattern of sandbag measures 33 inches × 14 inches empty, is made of canvas and issued in bales of 100, weighing 33 to 43 lbs.

When filled with about $\frac{1}{2}$ cubic foot of sand, &c., it measures over all about 20 inches × 10 inches × 5 inches, and weighs about 60 lbs.

Sandbag revetment is built at a slope of $\frac{4}{4}$ with alternate rows of headers and stretchers (the former with the chokes, the latter with the seams turned into the parapet), breaking joint (Pl. 2, Fig. 4). The revetment should end with a top course of headers. The bags must be laid *at right angles to the slope* and not horizontally or the revetment may slide. They should be not more than about three-quarters full, and should be beaten into the required shape when placed in position.

Heather, or scrub, made into bundles and built like sandbags makes a fair revetment.

Sandbags are chiefly used for loopholes or for repairing earth works, &c.

15. Sacks.

(*See Appendix III, Table* 11.)

1. Sacks or grain bags of which an army in the field usually has many thousands available, can be employed instead of sandbags and are easy to use. The rules for sandbag revetments apply also to sacks except that sacks should not be more than half full and may be

laid on all stretchers. It is not necessary to choke or tie up a sack if the mouth is carefully folded under it when it is being placed in position. The weight of the sack will prevent loss of earth.

The usual sizes will contain about 2 bushels (2½ cub. ft.) of grain, but should only be filled with about 1 cub. ft. (say 1 cwt.) of earth, &c.

16. Wire.

(See Appendix III, Table 14.)

1. Wire is principally used to form an obstacle either alone or in conjunction with abatis, and for lashing spars, &c.

The sizes, weights, &c., in normal use are given in Appendix III.

17. Canvas and Netting.

(See Appendix III, Table 11.)

1. Willesden canvas is kept as an article of store, in rolls up to 5 feet wide. For use as a revetment, stout pickets should be driven from 12 inches to 18 inches apart and anchored. The canvas can then be stretched between these and the parapet, and laced with wire to the top and bottom of the pickets. (Pl. 2, Fig. 6.)

It is useful for making weather-tight all forms of overhead cover and shelters, and for improvising stretchers for the removal of surplus earth from excavations, &c. It also forms a serviceable foundation for the support of a plank way to enable men to cross marshy ground. (Pl. 48, Fig. 9.)

2. Wire netting with stakes passed in and out of the meshes and anchored back forms a good revetment in soils which are not too sandy.

CHAPTER III.—WORKING PARTIES AND THEIR TASKS.

(See also Military Engineering, Pt. I, Sec. 4.)

18. GENERAL INSTRUCTIONS.

1. EXCEPT WHEN CIVIL LABOUR IS AVAILABLE, UNITS WILL AS FAR AS POSSIBLE CONSTRUCT THE ENTRENCHMENTS THEY ARE TO HOLD. (F.S. Regs., Pt. I, Sec. 5 (8).)

Engineer field companies will be employed to the best advantage under the orders of the divisional commander. They may be given separate tasks, such as the construction of obstacles of a technical nature, communications, observation posts, or the improvement of the water supply, &c., or be distributed to certain localities or sections of the defence. For instance, in the attack detachments of field companies may be employed with local reserves to open a way through obstacles, prepare captured localities for defence, and strengthen the main position, when gained, against counter-attack. (*See also* F.S. Regs., Pt. 1, Sec. 105 (v).) In the defence, it may be advisable to allot one or more sections of a field company to each section, to assist with the special technical knowledge and material at their disposal.

2. When making any but the smallest entrenchments, men are not kept continuously at work, but are changed at intervals. The total time is thus divided into periods called *reliefs*. As regards the length of reliefs a great deal depends upon the nature of the work, the total time it will take, the climate, and the condition of the men. This question of reliefs will also depend on whether the work has to be done at great speed, and whether it can be carried on by night as well as by day. For digging, short reliefs are best, and it will be found that a four hours relief is, as a rule, quite long enough for anyone but a trained sapper. With reliefs of two hours or less, time is wasted in handing over the work, tools, etc., from one party to another, unless the organization is very good.

3. With full-sized tools the average untrained soldier should excavate in ordinary easy soil the following volumes in each hour:—

1st hour	...	...	...	30 cubic feet.		
2nd „	...	...	...	25	„	„
3rd „	...	...	...	15	„	„
4th „	and after, up to 8 hours			10	„	„

or 80 cubic feet in a four hour relief.

If the soil is very easy these rates may be increased, and *vice versa*; and if two men are detailed to each set of tools these rates may be multiplied by $\frac{4}{3}$.

These rates hold good for a maximum horizontal throw of 12 feet, combined with a lift out of a trench 4 feet deep.

4. When the distance that the earth has to be thrown to deposit it in its final position is more than 12 feet, shovellers will be necessary as well as diggers.

5. The proportion of picks to shovels will be decided according to the nature of the soil. In ordinary easy soil the entrenching implement is almost equal to a pick for loosening ground. In moderately hard ground containing stones a good proportion of tools (excluding the entrenching implement) is found to be 110 shovels, 55 picks with 60 helves, and 10 crowbars per 100 men, working continuous reliefs up to a total of 40 hours. This proportion gives sufficient spare tools to enable defective ones to be rejected.

6. For continuous open trench-work, such as communication trenches, the normal distance apart at which men are spaced for work is two paces (5 feet). No practical increase of speed in execution can be obtained by reducing this spacing, unless it is not intended to make use of the full-sized pick. In this case men can work at 4 feet apart.

7. IN THE CASE OF FIRE TRENCHES (*see* SECTION 29) ALL QUESTIONS OF MECHANICAL SPACING AND DISTRIBUTION MUST GIVE WAY TO THE SELECTION, ON THE GROUND, OF THE BEST FIRING POINT FOR EACH AVAILABLE RIFLE.

These points must be entrenched first, and afterwards connected together as may be necessary for purposes of communication and command.

In normal broken ground each firing point, or small group of firing points forming a fire trench, must receive independent treatment both in design and execution in order to utilize to the utmost the existing facilities for good shooting, combined with concealment and protection. (Pl. 9, Fig. 5.)

8. Calculations cannot be profitably undertaken away from the actual sites of fire trenches except to give a general forecast of the amount of time and the distribution of labour required, *per rifle* of the fire trenches and cover trenches and *per pace* of the communication trenches, in accordance with the types of work selected *after reconnaissance of the actual ground*.

19. TASK WORK.

1. For instructional purposes task-work may be found better than time-work; but IN THE PRESENCE OF THE ENEMY MEN MUST WORK TO THE UTMOST LIMIT OF THEIR POWERS DURING THE WHOLE OF THE TIME FOR WHICH THEY ARE DETAILED FOR DUTY WITH A RELIEF.

Provision is made for the condition of the men by (*a*) adjustment of the length of each relief, and (*b*) by detailing numbers in excess of the tools available, as provided in Sections 18 (2) and (3) and 22 (1).

In calculating tasks it is better to underestimate than overestimate the men's powers, in order to avoid incomplete tasks.

2. In arranging tasks, the following rules should be adhered to:—

    (i) The task of the first relief should be larger than those of succeeding ones, as the diggers have less distance to lift the excavated earth.

    (ii) If possible, the men of each relief should leave a vertical face of earth for the next relief to commence upon.

    (iii) All diggers should commence on the left of their tasks, in order not to interfere with one another, and, in continuous trench-work, should break into the task on their right.

(iv) If not under fire, the earth first excavated should be thrown furthest away. (*See also* Sec. 29 (9).)

As the earth required for the parapet of a field work is obtained from the excavations (ditch and trench), the areas of the cross sections of parapet and of excavation must be approximately equal.

Figures shown thus ☐ 36 ☐ denote the approximate sectional area of the excavation or parapet in square feet.

The names of the different parts of a work in common use are illustrated in Pl. 6

20. ORGANIZATION OF WORKING PARTIES.

1. Care should be taken in arranging the preliminary details of working parties so that the men may arrive at the site of their work provided with tools, and in such formations as will admit of easy distribution to the works, which should be clearly traced in advance.

2. This tracing consists in laying out so much of the plan on the ground as is necessary to guide the distribution of the working parties. It may be done by spitlocking or by laying tapes, but in the case of fire trenches, pickets or stones should also, when possible, be used to indicate the chief line of fire for each rifle for the use of which cover is to be provided, as well as the width of the first task.

In hasty defence work tracing with a tape is usually only necessary for night work.

21. TOOL DEPÔTS.

1. During the execution of defence works temporary field depôts for tools and materials will be formed as close as possible in rear of each group of works in the section of the defence.

All tools not in use, and all materials not collected or delivered on the site of the works, or requiring special preparation, should be received, prepared or repaired, and issued at these depôts only.

2. Cooked food, water, shelter, and latrines for men of the working parties should be provided at each depôt, and messages intended for the superintendent of the works will be delivered to him there, or to the person detailed by him to take charge of the depôt and forward such messages.

3. The site of such a depôt may often correspond with the position of the shelters eventually provided for the supports to the fire trenches, as shown in Pl. 16, and it should be close to communications practicable for transport.

4. Before the arrival of the working parties, and on the conclusion of the works, tools will be laid out at the tool depôt, according to the detail of the several parties previously communicated to the person in charge of the depôt, either in rows or in heaps, the men in the former case filing on the rows and taking up a pick in the left hand and a shovel in the right, or filing between the heaps and receiving or depositing the tools as they pass.

Arrangements will be made that all tools issued from the tool depôt are sharp and serviceable.

In changing reliefs at the works it should not be necessary to collect the tools in use if the men of the new relief are properly detailed to relieve the men of the old relief individually, after previous observation and instruction, and if the relief is carried out gradually. (*See also* Section 28 (4).)

22. DISTRIBUTION OF WORKING PARTIES.

1. Working parties should be requisitioned in actual numbers of men required, and should be detailed from the same company, battalion, brigade, or division, and not employed in mixed detachments. In calculating their strength a reserve of one-tenth should be included. Unless otherwise ordered all ranks will proceed to their work fully armed, accoutred and rationed.

If the party is a large one and the work of a complicated nature, such as a redoubt, the men should be divided into detachments corresponding to definite portions of the work, and each detachment should be placed under an officer or N.C.O. well in advance of the time at which work is ordered to commence.

When a party is to work under the superintendence of an officer of another unit, the officer or N.C.O. in charge of the party will make himself acquainted with the work to be done and will himself direct the work of his party.

Superintendents should be relieved at different hours to the working parties, to ensure continuity in work.

2. Working parties should be distributed by one of the following methods :—

(i) Each party, having first extended to the required interval at a suitable distance in rear, is advanced and halted on the line of the proposed excavations, while an officer details each man to his task.

(ii) Each party is halted in column about 3 paces in rear of one flank of the proposed line of excavations, and formed in file or single rank according as one or two men are allotted to each task. The officer who has marked out the works then explains each task to the men as they arrive: the men moving off, wheeling to the right or left as the case may be, and forming up in succession on the alignment.

In either case a man, having marked out his task with a pick, places it on the left of his task, takes four paces to the rear, grounds arms, removes his accoutrements (*see* para. 5 below), and lies down till ordered to commence work.

3. No work must be commenced till the distribution of the whole of each party is complete, as it is difficult to remedy mistakes when work has once begun. The subsequent shifting of men invariably tends to confusion, loss of time, and possibly of tools, clothing and accoutrements

4. In cases where it has been impossible to reconnoitre the ground and actually site the works by daylight, and to avoid confusion and noise, it is better to divide up night working parties into squads and to construct short lengths of trench rather than to attempt to dig long

ones. The commander of each squad must see that the proper alignment is maintained, and should be provided with a tracing tape or rope divided into task lengths by pieces of white paper or cloth. He may also mark the ground with paper about 80 yards in front of the trench to ensure the lines of fire being kept low. For extending men for night work the second method described above is the better; but the normal interval may be slightly increased to prevent men striking each other.

5. If a sudden attack is likely to take place the rifles and accoutrements of all parties working in fire trenches must be within reach without necessitating the men leaving the cover of the trenches.

Superintendents, supernumerary numbers, and the reserves of working parties must be detailed to their exact posts and duties in case of a sudden emergency requiring the use of defence works in process of construction. This is especially necessary in savage warfare.

23. WORKING PARTY TABLE.

1. The following form may be useful to facilitate *rapid commencement of work* and to ensure that *men and tools are employed in the most advantageous manner*. It would, as a rule, only be applicable to cases in which it had been possible to send on officers in advance of the force to reconnoitre and report on the position to be held and upon the principal defensive works required.

It is not likely that more details would be supplied by the reconnoitring officers as to the works required than is given in column 1 on the form. It would then remain for the commander of the unit concerned (in this example a battalion) to detail his men and tools to the works in their respective order of importance and in the manner most likely to ensure the best use of the labour and tools available, as shown in Cols. 2, 3, 4 and 5. Should the tools with the unit not be sufficient, the commander would then apply to his immediate superior (in this case the brigadier) for the remainder.

The latter would then fill in Column 6, showing whence the balance of tools required was to be obtained.

No fixed times have been given for the completion of the works, as this will depend so much upon the tools available. Should any second reliefs be found necessary, a redistribution of tools and a fresh supply of labour can be arranged during the first relief.

2. In more deliberate work, it may be possible and necessary to forecast with more accuracy and detail the exact nature of the works, the number of men and tools required for their execution and the length of time needed. In such a case the form can be altered to suit the circumstances.

EXAMPLE OF WORKING PARTY TABLE.

WATLING RIDGE POSITION, No. 2 SECTION (13TH INFANTRY BRIGADE), 1ST SCOTS FUSILIERS.

1	2		3	4	5	6
Task. (1" Map Sheet 384.)	Men.		Tools required.	Tools with unit.	Balance to complete.	Remarks
	No	From				
1. East of HEXHAM COPSE at foot of slope (near rusty plough). Two 40-Rifle trenches, 18" command, traversed, recessed and with head cover. Soil easy. Will probably take 7 hours.	80	C Co.	40 picks 80 shovels	40 80	— —	
2. HEXHAM COPSE Clear. Brushwood and small trees. About 8,000 sq. yards. Some brushwood required for trench above.	90	A Co.	10 felling axes 70 billhooks or handaxes	10 49	— 21	*1st Dublin Fusiliers.*
3. West of NORTHAM FARM at foot of slope (cleft stick and paper). Two 30-Rifle trenches, 12" command. Soil very difficult. Probably take 8 hours	60	B Co.	60 picks 60 shovels	60 60	— —	
4. Communication trench from above trenches to east of NORTHAM FARM. 200 yards. Soil difficult.	120	D & E Cos.	120 picks 190 shovels	51 86	69 34	*use grubbers none available. 20 only Brig. Res.*
&c. &c.						

CHAPTER IV.—IMPROVEMENT OF THE FIELD OF FIRE AND UTILIZATION OF EXISTING COVER.

(See also F.S. Regs., Pt. I, Sec. 108, and Military Engineering, Pt. I, Sec. 10.)

For tools available and estimate of time and labour required, see tables Appendices I and II.

24. GENERAL INSTRUCTIONS.

1. The provision of a good field of fire will usually entail some clearance of the ground in front of the firing line, but this must be done in such a way as to give no assistance to the attackers in their advance or in the use of their weapons. At the same time the possibility of adapting and improving any existing cover for the use of defenders should be borne in mind. Natural obstacles, which may be left, should be such as not to interfere with counter attack or screen the enemy from fire.

2. It will be advisable first to improve the field of fire near the position and work forward as time permits; but, in case of a delaying action where fire effect at long ranges is required, it may be better to prepare for bringing fire to bear upon points at some distance from the position.

Hollows and unseen ground, which would give the attackers shelter at points dangerously near the position, may be filled up with abatis, or débris of walls, &c., and should be fenced off if possible.

Large scattered trees give less cover when standing than if cut down, and may sometimes be useful as range marks.

Range marks should be provided and should be placed on that side of large trees, houses, banks, &c., which is only visible to the defence. The simplest arrangement consists of one white object per 100 yards range. 500 yards may be denoted by the sign V, made with two boards, poles, &c., and 1,000 yards by the sign X. Intermediate hundreds being indicated by single objects in addition, as above described.

Every soldier should in addition know the ranges to points under fire from his post, which are likely to be traversed by the enemy. These points should not be selected merely because they are prominent.

3. Cover may be classed as cover from view, cover from fire, or a combination of both.

The chief uses of cover from view are to screen movements from the enemy's observation, with the object of effecting a surprise, and to enable supports and reserves to be moved as required with a minimum of loss.

When clearing the foreground of a defensive position, the question of leaving screens for this purpose should be constantly kept in mind. In a close country much can be done towards attaining this object by judicious thinning, or by leaving trees, hedgerows, or even portions of copses standing. Where no natural screens exist they may sometimes be improvised.

In utilizing existing cover to give protection from fire the essential point to bear in mind is that the soldier shall be able to use his rifle to the best advantage.

Cover, whether from view or fire, should not provide a good mark for the enemy's fire. The edges of woods, hedges and banks, which are clearly defined and run parallel to the enemy's fire position, prominent trees and other land marks, all present favourable objects on which to range, but the provision of alternative cover would often take longer to construct and be still more conspicuous.

4. Smoke may sometimes be employed to give cover for working parties, especially against search lights. Sacks filled rather tightly with straw, left open at each end and slit to allow the escape of the smoke, form simple and portable smoke producers. They should be lit in the centre of the straw, so as to burn outwards.

5. Thick brushwood, especially in the case of some tropical growths, forms a very effective obstacle, which should only be cleared away in accordance with the principles laid down above. Thus, in place of making a complete clearance, portions may with advantage be left untouched, either to deny special points to the attackers and break up their attack, or to compel the adoption of particular lines of advance. The portions cut down may often be formed into an obstacle among the parts left standing.

25. HEDGES.

1. Hedges and banks which interfere with the defenders' fire or screen the attack, must be removed so far as time will permit. The clearance of hedges perpendicular to the front is of less importance than of those parallel to it.

2. Ordinary hedges are principally valuable for the concealment they afford. Unless they are very thick, wire and stakes must be added to render them efficient obstacles.

3. Where there is a ditch on the defenders' side, it can usually be converted into a useful fire trench with little work. (Pl. 7, Fig. 4.) If there is no ditch on the defenders' side, a trench should be dug and the earth thrown up against the hedge, if command is necessary, but the hedge must be strong enough to support the earth and thick enough to prevent its showing through on the side nearest the enemy. (Pl. 7, Fig. 5.) The time required to excavate such trenches will usually be longer than that required for ordinary trenches on account of roots, and at first work must be concentrated only at the points to be occupied by each rifle-man.

26. EMBANKMENTS AND CUTTINGS.

1. Embankments are not, as a rule, good positions for a firing line exposed to artillery fire, as they offer opportunities for accurate ranging; they must, however, often be held in order to bring fire to bear on what

would otherwise be dead ground. They can be defended by occupying the rear side, as in Pl. 7, Fig. 1, or the front side, as in Fig. 2, or better still by a combination of both methods. The front side gives a better view of the ground, but cover can be obtained with less labour at the rear side.

2. Cuttings can be defended in a similar manner (Fig. 3). The rear side gives the best opportunity for concealing an obstacle; the front side is better for a subsequent advance, and secures good shelter for supports. Generally speaking fire trenches should be sited on the front side of cuttings at re-entering angles and on the rear side at salient angles, not necessarily on the actual edge of the cutting.

3. A road cut on the side of a hill will generally be visible to the artillery of the attack at long range, and should not therefore be held unless it offers special facilities for defence, or is artificially masked.

27. WALLS.

1. Walls can be knocked down by using picks, crowbars, and hammers; or a short length of rail slung from the pole of a limber, &c. Lightly constructed buildings may be similarly treated. If solidly built they must be blown down and the ruins levelled, as far as possible, so as not to give cover.

2. To give protection against rifle fire a wall must be well built and at least 9 inches thick. Walls should not be held under effective artillery fire, but may be utilized for defence after artillery fire has ceased. If it is desired to make use of a low wall and time is available, it should be used as a revetment, and a parapet thrown up against it.

A wall between 4 feet and 4 feet 6 inches high can be used as it stands. If a wall is less than 4 feet high, a small trench should be sunk on the inside to gain additional cover. (Pl. 8, Fig. 1.)

Between 5 feet and 6 feet in height, a wall can be notched; but, above 6 feet in height, a stage is necessary to enable men to fire over the wall, or through notches (Pl. 8, Figs. 2 and 3); or else the wall must be loopholed. (Pl. 8, Figs. 4 and 5.)

3. Loopholes, which are preferable to notches owing to the better headcover they give, should not be closer together than 3 feet from centre to centre and need not be spaced symmetrically. The opening can be made by means of crowbars or picks if a mason's chisel cannot be obtained, and should be as small as possible on the outside to lessen the chance of bullets entering. A rifle held in the required position will give the form and height of the loopholes. A normal type is illustrated in Pl. 10, Figs. 3 to 5.

4. Where long lines of wall have to be manned, flanking fire should, if possible, be provided from bullet-proof projections in front of the wall.

Care must be taken to avoid putting trenches unprovided with overhead cover so close in front of masonry walls that ricochet bullets or splinters of shells are liable to strike the occupants of the trenches.

5. The tops of walls, prepared for firing over, should be covered with a layer of turf to check splinters and provide a good rest for the rifle.

6. When a wall has been deliberately loopholed for fire as many dummy loopholes, or marks, as possible may be added to the exposed side, in order to multiply targets at close ranges. (*See also* Section 31 (4).)

CHAPTER V.—EARTHWORKS.

(*See also* F.S. Regs., Pt. I, Sec. 108, and Military Engineering, Pt. I, Sec. 8.)

28. GENERAL INSTRUCTIONS.

1. Earthworks may be classed generally under three heads, viz.:—Trenches, Redoubts, and Gun Emplacements.

Trenches are further distinguished as "fire trenches" or "cover trenches" according as they are for the firing line or for troops not actually engaged. "Communication trenches" are excavated covered ways connecting different parts of a position.

2 The value of concealment cannot be over-estimated, and every effort must be made to conceal the site of all earthworks in a position. It should be always borne in mind that invisibility is often as valuable as cover itself, while one carelessly constructed trench may give the enemy a good idea of the whole position.

The curves of parapets should be made to assimilate with the natural contour of the ground. Straight lines and sharp angles are, therefore, out of place. The fronts of parapets should be carefully covered with sods, transplanted bushes, &c., to make them resemble their surroundings. Cut branches become very conspicuous when withered, and if used should be changed by night.

3. If a parapet is placed on the sky line, spare earth may be piled up behind the trench and covered with turf, bushes, &c., to make a background for the defenders' heads, and to conceal its position. As a rule, however, a sky line is to be avoided. In this connection it must be borne in mind that the sky line will vary according to the spot from which the position is regarded. As it is usually most important that trenches, &c., should not be placed on what would be the sky line to the attacking artillery, the position should, if possible, be examined, whilst the siting of the trenches is under consideration, from the positions most likely to be occupied by the attacker's guns. It is not necessary that entrenchments should be of equal strength throughout a position. Such portions of the position as are liable to detection as the result of a distant examination by the enemy should receive earlier and more careful attention than those parts which will not be seen until he has drawn closer.

4. All earthworks, whether completed or not, must be concealed as far as possible, and all tools, materials, and signs of work in progress removed or concealed on each occasion that work proceeding in the presence, or possible presence, of the enemy, is suspended for a longer period than for a routine relief of the working parties.

5. Entrenchments will be used in attack as well as in defence. The main difference between the nature of the works constructed is that in the latter case the position will usually be selected and the work carried out more or less deliberately before fighting begins, while in the former all work will be carried out hastily on ground which may not have

been closely reconnoitred beforehand. (*See also* F.S. Regs., Pt. 1, Sec. 105.)

6. While the principles of the tactical employment of earthworks must always be borne in mind, the works illustrated in the plates should be regarded as types only and should be varied to suit local conditions, every effort being made to save time, labour and material by utilizing and improving existing cover.

29. FIRE TRENCHES.

1. The ideal site for a trench is one from which the best fire effect can be obtained, in combination with complete concealment of the trench, and of the movements of supports and reserves in rear. Such positions being rarely found, the best compromise must be sought, bearing in mind that a good field of fire up to about 400 yards is of primary importance.

2. When the position includes commanding ground the firing line need not necessarily be on it. The advantage of high ground for a defensive position is often over-estimated. It is, however, desirable that the position should conceal and shelter the defender's reserves and communications, while enabling the movements of the enemy to be observed.

3. It may sometimes be advisable to place the infantry fire trenches at or near the foot of a slope so as to obtain a grazing fire, while the artillery is posted on higher ground in rear. It must, however, be remembered that it will be difficult, if not impossible, to reinforce the defenders of such trenches, or to supply them with ammunition, water, food, etc., during daylight.

4. PROVIDED THE FIELD OF FIRE IS GOOD, A PARAPET CANNOT BE TOO LOW, AND IN SOME CASES NO PARAPET AT ALL NEED BE PROVIDED. Every endeavour should be made to arrange the trenches so that the front of one is swept by the fire from those on either hand, for which purpose short trenches up to 40 yards or so in length are more easily adapted to the ground than those of greater length. (Pl. 18.)

Earth which is not required should be carried away to some spot under cover, or formed into dummy parapets. If wheelbarrows are not available, earth may be carried away in sandbags or in squares of matting, etc., slung to a pole.

5. Every artifice should be used to mislead the enemy as to the positions of the trenches and guns, *e.g.*, conspicuous dummy parapets, not in the alignment of either real fire trenches, or of closely supporting artillery may be thrown up to draw his fire, and may also be equipped with dummy guns, paper masks, helmets, &c. They would be specially suitable if used in conjunction with a false or advanced position. (*See* Sec. 50 (10).)

Scrub, long grass, etc., forming a natural screen to trenches should not be trampled down or otherwise interfered with more than is absolutely necessary to give a clear field of fire.

6. Turf which may be needed should be taken from some unseen spot, or it may be possible to take it from a strip of ground, which with a little labour may be made to resemble a trench. Turf used for

concealing parapets should be laid so that spaces do not occur between adjacent sods.

7. The *design* of a trench will depend on the time and labour available, on the soil, on the site, and on the range and description of fire which may be brought to bear on it, but the following rules are common to all:—

(1) The parapet should be bullet proof at the top.

(2) The parapet and trench should be as inconspicuous as possible.

(3) The interior slope should be as steep as possible.

(4) The trench should usually be wide enough to admit of the passage of a stretcher without interfering with the men firing, and if a step is provided as a banquette, it should not exceed 18 inches in width (Pl. 9, Fig. 3).

(5) The interior should be protected, as far as possible, against oblique and enfilade fire, and from reverse fire if there is a danger of fire coming from the rear.

(6) Arrangements for drainage should be made.

8. Types of fire trenches are given in Pl. 9.

To excavate a length of 2 paces per digger of a trench of the type shown in Fig. 1, will take an untrained man about $1\frac{1}{2}$ hours, in moderately easy ground. This does not allow for providing an elbow rest or concealing the parapet.

Should time be available, cover and facility of communication may be much improved by deepening and widening the trench, as shown on Pl. 9, Fig. 3, which allows room for men to pass behind the firing line without disturbing it.

Should a higher command than 1 foot 6 inches be required to enable the defenders to see the ground in front, the parapet must be heightened with earth obtained by widening and deepening the trench. A firing step, which should not exceed $1\frac{1}{2}$ feet wide, unless overhead cover is provided, is necessary $4\frac{1}{2}$ feet below the top of the parapet; the interior slope of this step must be revetted.

Pl. 9, Fig. 2, is a case where the ground in front can be seen without any command. For the sake of concealment the excavated earth must be removed or formed into a dummy parapet elsewhere, and the back inside edge of the trench covered up with grass, sods, etc.

9. When entrenching under fire, or in the attack, each man should provide himself with cover as quickly as possible. Pl. 9, Figs. 4 and 5, shows how fairly good cover can be rapidly obtained for individual men lying down. Sods and lumps of earth should be used to revet the inner slope of the parapet. If more time is available the trenches can be improved and deepened as shown by the dotted lines.

In constructing fire trenches in the presence of the enemy, care must be taken to utilize the earth as it is excavated so as gradually to improve the cover provided. It must not be thrown haphazard to the front with a view to subsequent arrangement.

10. An elbow rest is useful because it supports the arm while firing, and is convenient for ammunition, but it is wasteful of head cover, and the vertical exposure of the firer is greater than when no elbow rest is used. It should be 9 inches below the crest and 18 inches wide.

11. Fire trenches should usually be provided with small recesses in which to place packets of ammunition. These will also serve as steps by which to reach the crest, should an advance be ordered. (Pl. 9, Fig. 2.)

12. Ramming earth decreases its resistance to bullets; it should, therefore, be allowed to lie naturally as thrown up, except in the case of shelters constructed under a parapet (Pl. 12, Fig. 3), when some ramming may be advisable to prevent water percolating into them.

30. ENTRENCHING IN FROZEN GROUND.

1. If it is necessary to entrench ground which is frozen, and if there is ample time, the following method will save the heavy labour otherwise required.

A layer of straw 12 to 20 inches thick should be spread so as to rather more than cover the area to be excavated. The straw, having first been covered by a thin layer of earth, is then set on fire at intervals of about 5 yards. The burning should be allowed to go on for 12 hours before the ashes are removed, and digging commences.

2. If water is poured over a parapet constructed during frost its resistance to rifle fire will be increased.

3. The sound of digging in frozen ground can be heard at a distance of about half a mile, while the sparks caused by picks striking stones have been seen up to about 600 yards.

31. LOOPHOLES.

1. Head cover necessitates the provision of some form of loophole or notch but tends to diminish the number of rifles that can be put in line, as well as to reduce the field of fire and view. It generally makes the work more conspicuous, but is of undoubted advantage owing to the feeling of security it inspires, and consequent greater accuracy of fire.

2. Careful arrangement is necessary to ensure the maximum of fire effect and invisibility.

Invisibility is the first consideration during the early stages of the battle and at long ranges, while cover increases in importance as ranges diminish; it may, therefore, occasionally be advisable not to construct head cover at first, but to have materials, such as filled sandbags, ready in the trenches to enable this to be done when further concealment is useless.

3. Loopholes can be made of sandbags, sods, or various other materials such as boxes or sacks filled with earth, or gravel.

The size and shape of the opening must be governed by the extent of ground to be covered by fire, but it will also depend on the width of the parapet, which will vary according to the resisting power of the material used in its construction.

The minimum height of openings for a parapet 3 feet 6 inches thick on *level ground*, using the service rifle at 2,000 yards range, is, for the inside, six inches; for the outside, four inches. The interior height of an earth loophole should seldom be less than 14 inches. This additional allowance is made to enable men to get their heads well forward and under the head cover.

Each loophole must be tested with a rifle, with the bolt removed, to ensure that neither the line of sight nor the line of fire are obstructed.

4. Loopholes should always be blinded by means of a ball of grass, straw or leaves, until they are actually manned. The shadow thrown by the loophole, as well as the hole itself, may be masked by light screens of branches, heather, dry grass, etc. (Pl. 11, Figs. 1 and 2), but it is little use concealing the loopholes if the parapet itself is conspicuous.

5. If the trench is sited for firing downhill, the parapet must be sloped off to the lowest line of fire, before commencing the loopholes. This principle should be equally observed in the case of parapets in elevated positions, or when it is required to fire uphill.

6. Pl. 10, Fig. 1, shows the three types (A, B and C) of loophole which are usually employed.

Loopholes of type A give the defender a good view, and enable him to fire on any point within the angle of opening without moving his position to any great extent. This type of loophole is difficult to conceal, and, if covered in, the top sandbags require strong support to prevent them sagging. (Pl. 10, Fig. 2.)

Loopholes of type B are less visible to the enemy, but they are more difficult to observe from and fire through. They are suitable for use in masonry walls, &c. (Pl. 10, Figs. 3 to 5.)

Loopholes of type C (Pl. 11, Fig. 3) are a compromise between the first two types and can be adapted to give a large field of fire. Thickening the parapet reduces the area round the loophole which can be penetrated by bullets. This is not the case with the first two types.

7. If there is a supply of shingle or gravel available, loopholes may be made as follows:—Place one sandbag inside another, and fill the inner one with the shingle. Three of these double sandbags will make a loophole, and the rest of the headcover can be revetted with single sandbags filled with stones. The sandbags forming the loophole should be placed on edge and stamped or hammered out until the opening is curved as shown, and the bag is about 9″ wide and 6″ high. This loophole requires comparatively few sandbags, and the parapet does not require any extra thickness of earth, while the loopholes can be as close as 3′ 3″. (Pl. 11, Figs. 1 and 2.)

8. A form of loophole which has the advantage of giving a wide field of fire and view is a continuous slit along the parapet, except for the supports required for the material above. (Pl. 25, Figs. 2 and 3.)

9. Steel loophole plates are articles of store (Pl. 10, Fig. 6). They may be arranged as shown on Pl. 11, Fig. 3. They make the best head cover, as they are bullet proof, but they would rarely be available for hasty defence work. In the vicinity of railways a strong steel loophole can be made with a string of fish-plates threaded on a rod.

32. DRAINAGE OF TRENCHES.

1. The drainage of trenches must be attended to from the first. The bottom of a trench should be sloped to a gutter, which should preferably be made along the back of the trench. Any water collecting in it should be led off to lower ground, or else into soak pits, which may be about 2 or 3 feet in diameter and 3 feet deep, and filled with large stones.

Care must be taken to prevent rain water running into the trenches from the surrounding ground.

33. TRAVERSES.

1. Trenches exposed to enfilade fire and to the oblique fire of artillery, should be traversed and recessed. Traverses give protection against enfilade fire, and also localize the effect of a shell bursting in the trench. (Pl. 12, Figs. 1, 2 and 3.) It is better to make several small traverses than one or two large ones. When the ground is suitable, an irregular line of trench may obviate the construction of traverses, but the best lines of fire must never be sacrificed for this reason. Against oblique or enfilade fire from long ranges, traverses alone will not suffice, on account of the steep angle of descent of the bullets, and overhead cover may be necessary. Recesses in the parapet, large enough to hold one or two men, give protection against such fire, but seriously reduce the number of rifles that can be employed (Pl. 13, Fig. 1). Such recesses are best made after the trench has been excavāted.

34. COMMUNICATION TRENCHES.

1. If time admits, covered communications must be arranged from the firing line to the rear (Pl. 16). These, by concealing the movements of the defenders, will permit of the firing line being reduced to a minimum in cases where it is being attacked by artillery fire alone, or where the attacking infantry is out of range, and will also enable the supports to reach the firing line under cover. A trench similar to Pl. 12, Fig. 4, will usually suffice.

Time and labour in the construction of these trenches will be economized by a skilful use of the ground and by reducing the distance between the cover for the supports and firing line as much as possible.

They may require parapets on both sides, and where exposed to view or enfilade fire should be traversed and given overhead cover. (Pl. 16.)

2. In positions where their employment as fire trenches might subsequently be desirable, trenches similar to Pl. 9, Fig. 1, may be used.

3. For advancing towards the enemy by means of covered communications leading from the firing line *to the front, see* Mil. Eng., Part II, Secs. 4 and 5.

35. COVER TRENCHES.

Cover trenches are useful to protect any men who are not using their rifles. When time is limited and materials are not at hand, a section similar to Pl. 9, Fig. 1, but with slightly higher parapet and no elbow rest, may be employed. If more time and material be available, trenches similar to those shown on Pl. 9, Fig. 3 and Pl. 20, Figs. 1 and 2, should be used.

36. OVERHEAD COVER AND SHELTERS.

1. Overhead cover requires a large amount of material, which will often not be available, and takes a long time to construct. Its main

advantage is that it enables the defenders of a trench to continue to use their rifles, even when exposed to a heavy shrapnel fire. Its importance will increase with the progress of aviation and the use of short range grenades and bombs, and its employment in some parts of the firing line may, therefore, be desirable.

2. Overhead cover to keep out splinters of shells, shrapnel bullets and hand grenades, should consist of about 9 to 12 inches of earth, or about 3 inches of shingle, supported on brushwood, boards, corrugated iron, or other material.

3. In constructing splinter-proof shelters in the firing line it should be recollected that:—

    (*a*) the parapet must not be unduly weakened by them ;

    (*b*) they must not curtail the number of rifles available ;

    (*c*) it must be possible to get in and out of them quickly ;

    (*d*) simple and numerous shelters are better than a few elaborate ones.

4. Various forms of splinter-proof shelter are shown on Plates 12, 13, 20 and 21. They all require a great deal of material. They should always be given transverse partitions to localize the effect of shell and be made weather-tight if possible. In firm soil it may be possible to burrow or tunnel shelters in reverse slopes which will suffice without additional materials.

5. In the case of closed works when artillery attack is expected from the front only, splinter-proof shelter for men allotted to the flank defences, may be given in trenches roughly parallel to the front faces. The trenches may be continued with advantage across the whole redoubt, for purposes of communication.

When the artillery attack may come from any direction, shelters for men not firing must be arranged to face in different directions, and parados must be constructed. (Pl. 12, Fig. 3 ; Pl. 19, Figs. 2 and 3 ; Pl. 20, Fig. 1.)

6. A roof proof against all except heavy howitzer shell can be made with two layers of rails parallel to the chief line of fire, falling to the rear at a slope of $\frac{1}{4}$ or steeper, and separated by 1 foot of earth, the top layer being covered with 2 feet of earth.

If timber only is available, the roof may consist of 12-inch logs, with 7 feet of earth above.

7. Where overhead cover is provided arrangements should, if possible, be made for extra rifles to fire over the top of the cover at night or upon emergency. (Pl. 13, Fig. 4.)

37. Protected Look-outs and Observation Posts.

In connection with all fire trenches protected look-outs should, if possible, be provided, which should be indistinguishable from the front. Well-made loop-holes may be sufficient for the purpose. (Pl. 13, Fig. 4.) Double reflecting mirrors, as provided under Musketry Regns., Pt. II, para. 185, may be usefully employed when obtainable.

Artillery and other commanders will also require observation posts from which to maintain a general control over the fire. Such posts will often be placed clear of all trenches and emplacements, wherever an extensive view is obtainable. The post for sentry group shown on Pl. 30, Fig. 2, can be readily adapted to such a purpose, or it may be necessary to provide raised observatories. (Pl. 45, Fig. 1.)

38. DRESSING STATIONS.

Some covered dressing stations should always be prepared in rear of the fire trenches. Each should be large enough to contain a plank table 6ft. 6in. × 2ft. 6in. with 2ft. clear space all round. These stations should be placed near any cover trenches or splinter-proofs that may have been made in order that these may serve as waiting rooms for the wounded, otherwise the communication trenches are liable to become blocked. (Pls. 16 and 17.)

39. COVER FOR ARTILLERY.

(*See also* F.S. Regs., Pt. I, Secs. 105 and 108.)

1. Cover from both fire and view may be obtained by a combination of natural and artificial cover. Thus the reverse slope of a ridge may be scarped, or pits may be sunk close to the front crest of a feature, but, unless artificial platforms are available, it is generally inadvisable to disturb the natural surface of the ground, as it does not cut up so readily as ground which has once been broken. It is better to take advantage of natural screens and to construct epaulments in rear of them.

2. If the ground is dry it may be watered or covered with raw hides to prevent the guns being located by the dust which would otherwise be thrown up on discharge.

3. The primary object is to supplement the cover afforded by the shield sufficiently to protect the detachments against oblique fire or against shell which strike the ground just short of the shield.

4. Diagrams of a gun pit and epaulment, suitable for shielded guns, are given on Pl. 14. *These types are given as a guide only*, and must be modified to suit the ground. For heavy guns, the epaulment should be constructed on the same principle as for field guns, but a parapet about 8 ft. high should extend round the front of the carriage.

The parapet should be constructed so as to provide ample interior space and a wide field of fire, and the gun should be run up as close to the parapet as possible.

Each field gun requires for its working a breadth of 18 ft. between the detachment trenches.

5. The original trace of the pit or epaulment should admit of improvement in the event of more time becoming available.

With more time the following should be added :

    i. Alternative emplacements.

    ii. Trenches to cover the detachment standing and recesses for ammunition.

    iii. Traverses against enfilade fire, and overhead cover against howitzer shrapnel bullets.

    iv. Cover for an unlimbered wagon near each gun with a covered approach for men replenishing ammunition; or some shelter pits for men bringing up ammunition from the wagon line.

40. Cover for Machine Guns.

(See also F. S. Regs., Pt. I, Secs. 7 and 109.)

1. The value of machine guns largely depends on their being at hand when favourable opportunities for their employment occur. In the attack such positions will be gained almost entirely by a skilful use of the ground.

2. In the defence, on the other hand, concealed emplacements can be prepared in anticipation. These sites should usually be selected with a view to bringing a powerful enfilade or oblique fire on the attackers after they have reached effective infantry range, to flanking supporting works, and to sweeping any gaps that may have been left in the line of obstacles. (Plate 18.)

3. Machine guns can fire over a height of from $14\frac{1}{2}$ to 30 inches. In selecting a site for a machine gun emplacement particular attention should be paid to concealment. Provided the probable lines of advance of the enemy can be swept with fire, it will usually be preferable to select positions for machine guns away from salients, works, villages, etc., that may be held, so that they may be free to move about.

4. If the machine gun is to be located in a trench a platform of earth at the requisite depth can be left as the trench is being dug, or it can be built up subsequently. The crest of such an emplacement may take the form of the arc of a circle (Pl. 15, Fig. 3), the length of which will depend on the extent of ground it is desired to sweep with fire. Headcover should, if possible, be provided, but must not appear different from that constructed elsewhere in the trench, and the height and form of the openings through which the gun is to fire must be regulated by actual trial over the gun sights. The front of the emplacement may be undercut to take the front legs of the tripod, should the nature of the soil permit. Two types of emplacement for a machine gun are shown on Pl. 15.

5. Splinter-proof shelter for the detachments should be provided close to the emplacements.

41. Field Redoubts and Closed Groups of Trenches.

1. Field redoubts are works entirely enclosed by defensible parapets which give all-round rifle fire, and may be of any suitable command. This all-round defence gives them greater resisting power against infantry than a group of trenches, and minimizes the danger of the local penetration of a defence zone by the enemy, especially at night.

2. UNDER ORDINARY CONDITIONS REDOUBTS IN DEFENSIVE POSITIONS MUST NOT BE DESIGNED OR SITED IN SUCH A WAY THAT THEY CAN BE RECOGNIZED AS SUCH BY THE ENEMY, since this would probably lead to such a concentration of artillery fire on them as to neutralize their special value. This will prevent their employment in the main zone of defence as a general rule. If, notwithstanding, a redoubt is required in such a position inconspicuousness becomes a very important consideration. This will usually entail a command similar to that of the neighbouring fire trenches, combined, when time permits, with shelters

in rear for the troops not actually engaged, an efficient obstacle, and the various auxiliary provisions requisite to render it self-contained and independent of adjacent works as regards capture by direct assault. Such an arrangement of fire trenches is shown on Plates 17 and 19.

In order to take its place in the main zone of defence such a group should, in addition, be planned so as to be elastic in its fire capacity while still remaining self-contained. It will thus be possible to vary the garrison of such a group, either with a view to swelling the numbers in a general forward movement, or to correspond with the momentary fire value of that section of an entrenched zone in which it lies.

3. Redoubts may, also, often be employed as supporting points in rear. In such retired positions there will generally be sites which, while commanding a good field of fire, will not be visible from a distance.

4. For a work placed as a supporting point behind the front line, which is not visible from the attacker's artillery positions, the parapet may be higher. A high command has three advantages:—

    (1) It has a better command of its field of fire than a low redoubt.
    (2) It has a better moral effect on its defenders.
    (3) It conceals the interior of the redoubt from view.

The disadvantages lie in the extra labour and time entailed in making the parapet. A type of redoubt with high command is shown in Pl. 21.

5. Redoubts may be used for detached posts, and posts on lines of communication. Such works will often have to be a refuge and depôt for passing troops, and room inside must be allowed for this purpose. It will hardly be possible to make these works invisible, as it is essential that the parapets should be high enough to allow the defenders to move about the interior without being seen. Being isolated and conspicuous, they may be liable to bombardment, and a large amount of splinter-proof shelter should be built. An all-round obstacle for protection against night attacks should be provided.

The site should be such that the immediate foreground may be well swept by the fire from the parapet, and the work should be so disposed as to give the strongest possible fire on the enemy's best lines of attack. There must be no dead ground at the angles.

6. The plan or trace of a redoubt or closed group will depend on—

    (a) Fire effect required from it.
    (b) Configuration of the ground.
    (c) Probable variation of garrison.

7. There is no necessity for symmetry in the design of a redoubt, although it has advantages. On a level site a rectangle with blunted angles is a suitable form to employ. Faces which make a considerable angle with their neighbours, as in a rectangle, should not be less than 20 yards in length, and the short faces which blunt the angles should be at least 10 yards long in order to give room for an effective fire from them. For the same reason, curved faces, which it is often convenient to use, should be struck with a radius of not less than 20 yards. A

complete circle should be avoided, except for very small posts, as its fire is weak in every direction. In the case of a closed group of fire trenches symmetry will, as a rule, be out of the question, and considerations (a) (b) (c) above will determine the trace.

8. The maximum garrison should always consist of one or more complete units. The rules laid down in F.S. Regs., Pt. I, Sec. 108 (8), for calculating the number of troops required for the defence of a position should be followed when making estimates of the numbers required for the defence of a redoubt or group.

9. Since a redoubt is intended for all-round defence, precautions must be taken to prevent the defenders suffering from reverse fire.

10. The entrance to a redoubt used in civilized war may be a gap left in the face least exposed to attack, but under the close fire of the defence. The entrance should be wide enough to admit a wagon, i.e., not less than 8 feet.

11. Good drainage of the redoubt and shelters must always be provided, and should be put in hand as soon as the work is commenced. Soak pits will seldom suffice for this purpose, and the drains should, if possible, be led out of the redoubt to lower ground.

12. When a redoubt or fire trench is to be occupied for more than a few hours, latrines and cooking-places and storage for water, accessible by covered communication, should be provided. (Pls. 16, 17, and 19, Fig. 1; and Pl. 20, Figs. 8 and 4.)

18. As redoubts are intended to withstand assault they should always be provided with obstacles, which should be placed where the fire of the defenders will cover them most effectively.

The nearer the obstacle is to the parapet the less elaborate need it be and the less labour and material will its construction demand; while its defence can be more effectively carried out, especially at night. For further details regarding obstacles, see Ch. VI.

CHAPTER VI.—OBSTACLES.

(*See also* Military Engineering, Pt. I, Sec. 11.)

For table of time, men, and tools required, see Appendix II.

42. GENERAL INSTRUCTIONS.

1. OBSTACLES ARE USED TO OBTAIN A DEFINITE CONTROL, BOTH AS REGARDS DIRECTION AND SPEED, OVER THE PROGRESS OF TROOPS ADVANCING TO THE ATTACK. THEIR CHIEF VALUE LIES IN THEIR POWER TO DEFLECT THE ATTACKING TROOPS INTO AREAS MOST FAVOURABLE FOR THEIR DESTRUCTION BY THE DEFENDERS. With this object in view they should be arranged (1) to break up the unity of action and cohesion of the attacking troops, (2) to deflect the parties thus isolated into the best swept fields of fire, and (3) to arrest them under the close fire of the defenders. They are especially useful against night attacks.

2. For the first two purposes indicated above, the obstacles may be at any distance from the defence, but, if they are not under fire from the position, much time and labour will be needed in order that they may effect their object.

For the last purpose, they should fulfil the following conditions:—

(a) They should be under the close rifle fire of the defender, the outer edge not more than about 100 yards from the parapet. For small posts or redoubts they should be *quite close*, so that they may be effectively defended at night; but, if they are much less than the distance given above, it will be possible to throw hand grenades into the work. They should be as wide as time and material will allow, should afford the enemy no cover, and should, if possible, be sheltered from his artillery fire. Their actual position will generally be determined by placing them where they can be covered by the most effective fire of the defenders.

(b) They should be difficult to remove or surmount, and will be most effective if special appliances, not usually carried by troops, are required for their removal. Special attention should be paid to the security of their anchorages.

(c) They should, if possible, be so placed that their exact position and nature are unknown to the attacking force ; with this object their sites may be sunk. It must also be remembered that conspicuously placed obstacles may betray the existence of an otherwise well concealed position.

(d) They should be arranged so as not to impede counter attacks.

(e) They should not be constructed without authority from the officer commanding the section of the defence, otherwise they may interfere with an advance.

(f) They should not be continuous, but should be constructed in sections. Occasional gaps in the line will often lead the attackers to crowd in towards them. Such passages may be provided with land mines, and must be covered by gun and rifle fire. Roads passing through obstacles, and occasionally required for use by the defence, should be closed by portable obstacles, such as chevaux de frise, when not required for traffic.

3. The greatest length of obstacle which can be controlled by one man on a stormy and dark night is about 35 yards on either side. Where the total length of obstacle exceeds 70 yards, additional sentries, systematic patrolling, or efficient mechanical alarm signals (which will act upon interference not amounting to the severing of a wire) must be provided in order that the defence may reap the full benefit which an obstacle should confer.

4. The provision of obstacles, flares, and mechanical signals is under no circumstances to be accepted as a substitute for the efficient performance of the duties of a sentry as laid down in F.S. Regns., Pt. I, Section 81.

43. ABATIS.

1. The construction of abatis can seldom be attempted unless suitable material grows close at hand. Abatis formed of limbs of trees firmly picketed down with the branches pointed and turned towards the enemy form a very efficient obstacle. Strands of wire interlaced between the branches add still further to their efficiency.

2. In uncivilized warfare where bush or jungle abounds abatis of thorn bushes form an efficient obstacle. It should be made about 4 feet high and not less than 15 ft. wide.

3. Tree entanglements are formed by cutting trees, brushwood, &c., nearly through at a height of about 3 feet, and securing the branches by pickets to the ground. They can often be constructed whilst clearing the foreground, and may be suitably employed at the edges of woods and orchards or for blocking roads.

4. Timber which is felled within close rifle range should usually be cut off breast high so that the stumps may be available for wire entanglement.

44. WIRE OBSTACLES.

1. A low wire entanglement can sometimes be usefully employed in undergrowth. It can be made by driving pickets into the ground, so as to leave 12 or 18 inches projecting. The tops of these pickets are then joined to one another with wire.

2. High wire entanglement forms a very effective obstacle. It should be as wide as time and material will allow, but the width need not be uniform. It should be greatest where the fire of defence is least effective. It will be more difficult to remove if it is constructed in two zones with a small space between the parts.

8. The stouter the posts that are used, the better; but, if they are very large, it will be difficult to drive them into the ground, while their stability suffers if holes have to be dug for them. They should average 4 to 5 inches in diameter, and 5 to 8 feet in length.

4. The outside posts must be well stayed, especially those nearest the defenders, and, to render the passage of the obstacle by means of hurdles, planks, &c., more difficult, the posts should be driven in at irregular intervals (5 to 8 feet), and to varying heights (average 4 feet). With the same object, large nails may be driven into the heads of the posts.

5. Each post should be joined with taut wire, head to foot and foot to head, to all the adjoining ones. The wire should be wound round the posts and secured by staples, which may be made of the wire itself. Barbed wire should then be hung in festoons between the posts, but on no fixed pattern, and fastened to the posts by short lengths of wire.

6. Pl. 22, Figs. 1 and 2, show the principles of construction of a high wire entanglement, while Fig. 3 shows a method of improving a wire fence into a fair obstacle.

7. IT IS ESSENTIAL TO THE EFFICIENCY OF THE OBSTACLE THAT IT SHOULD BE IMPOSSIBLE TO CRAWL UNDER IT WITHOUT THE USE OF CUTTING TOOLS.

8. To ensure something being done at once and throughout, it may be advisable to order the construction of a fence of so many strands between the selected points which will form a nucleus for the finished obstacle.

9. Trip wires may be put in front of likely points of attack. They should be in lengths of about 10 yards, fastened at each end to a stout peg, hammered flush with the ground. The wire should be quite loose, and should be tightly coiled up before use so that it may be curly when placed in position.

45. BARRICADES.

1. Barricades to close streets, roads and bridges, can be made of almost any materials. They should not, as a rule, completely close the road to traffic, but should be made in two overlapping portions, or be placed where a house standing back from the general line of building allows a passage round them.

2. If the defenders are to fire over them, they should be bullet proof, and, if placed in a street, they should be flanked front and rear by fire from adjacent houses. (Pl. 24, Fig. 2.) For firing four ways they should be arranged as shown in Pl. 25, Fig. 3.

46. MINES.

1. Fougasses and land mines have great moral, and sometimes material, effect on the attackers, but are difficult to construct and should only be laid by officers who have a thorough knowledge of explosives (see Military Engineering, Part IV).

47. INUNDATIONS.

1. Inundations can be made by damming up a stream. A bridge is a good place to select for the purpose. If the dam is made of earth

either loose or in sacks, it is essential that a waste weir or overflow, cut in the solid ground clear of one end of the dam, should be provided, or the destruction of the dam will rapidly follow its completion. An earth dam must have a gentle slope, not steeper than ⅓, next the water.

2. If the inundation is likely to be shallow, the ground should be first prepared by digging irregular trenches and holes, or one regular trench which will make the water too deep for wading, and by the construction of wire entanglements; this will render the passage of even a very shallow inundation a difficult matter.

48. PASSAGE OF OBSTACLES.

1. Obstacles such as wire entanglements suffer very little from artillery fire or from explosives thrown indiscriminately into them. All obstacles may be passed by bridging over, cutting through, or tunnelling under, but the operation must, as a rule, be a surprise, and should generally be carried out in darkness. Whatever plan is adopted it must be capable of being carried out rapidly and silently, must be free from complicated manœuvres, and must be preceded by a very careful reconnaissance of the actual sites selected for passage.

2. The following are some of the methods that may be employed for passing wire entanglements:—

(i) By climbing over the entanglement on specially prepared ways. These may be from 9 feet to 10 feet long and from 2 feet to 3 feet 6 inches broad, made of light battens nailed to longitudinals (say 1½ inches by 5 inches) with rounded ends and bottoms. These frames may be placed in position by up-ending or sliding them along the wires; or, if connected by flexible joints (rope), they may be carried folded up and unfolded on the top of the entanglement. Planks, hurdles or ladders may be used in a similar manner, but a narrow gangway is easily capsized.

(ii) By clearing a passage through the entanglement:—
   (a) By the deliberate application of explosives. One way of doing this is to lash a charge on to a long pole, or to fill canvas hose, etc., with it and push it along the ground under the wires. If the depth of the entanglement is too wide for one pole the number must be increased, and they must be placed in position by men crawling along under the wires. Care must be taken that the charges on the different poles overlap so as to ensure the continuity of the whole. Two pounds per foot run of width should clear a passage of from 10 to 15 feet wide, but this result is not always attained. The objections to this method are the excessive quantity of explosive required, the difficulty of placing such a charge, and the fact that when the charge explodes all chance of a surprise is at an end.
   (b) By cutting the wires. This would be done by men creeping up to the entanglement and cutting the wires. To prevent noise when a wire is cut a man should hold the wire on either side of the cutter.

(c) By using grapnels to pull a piece of the entanglement away. Grapnels of steel of about ¾-inch diameter should be used, with 3 or 4 short prongs and an eye at the end to take a wire rope. They should not exceed 4 lbs. in weight, as otherwise they are too heavy to throw. When they have caught in the entanglement, the latter would be dragged away bodily by men hauling on the rope. This method would usually be combined with (b)

3. For the passage of obstacles requiring the further application of bridging or explosives, *see* Chapters XIII and XIV, and for mining or tunnelling underground, *see* Military Engineering, Part IV.

49. WIRES AND LEADS.

In dealing with wires which may be connected to alarm signals, mines, &c., care should be taken not to vary their tension during or after their severance. If the wires are supported by staples the latter should be driven tight home into the posts on either side before making a cut, and the severed ends carefully twisted round the posts to maintain the tension.

If the wires are covered with insulating material, the cut ends should not be allowed to touch other bare wires or the earth, but should be carefully supported in the air as far apart as possible.

CHAPTER VII.—DEFENCE OF LOCALITIES AND POSITIONS.

50. Defence of an Extended Position.

(*See also* F.S. Regs., Pt. I, Sec. 108.)

1. Every endeavour must be made to follow the principles for an active defence given in the Field Service Regulations, and especially in Pt. I, Section 108 (2) where it is provided that "Though the extent of ground actually held, when the direction of the enemy's advance is definitely known, must be strictly limited by the numbers available, the extent of ground reconnoitred and prepared for occupation may be much larger, and should admit of various alternative distributions of the force to meet the various courses of action open to the enemy."

The preparation of alternative firing points, and improvements to communications should therefore usually take precedence of other work as soon as the main tactical points have received that amount of attention which the commander of an entrenched zone may consider reasonably sufficient.

2. The object with which a position is occupied as well as the time likely to be available for its preparation will usually determine the works to be undertaken, but, as no position can ever be too strong, additions and improvements should continue to be made up to the last moment. Such improvements are not to be confined to the provision of better cover from fire for the defenders. Special consideration must be given to the means by which the artillery will co-operate with the infantry.

3. In practically all positions there will be localities of special tactical importance. It is on localities of this nature that the principal efforts of the defender should be concentrated in order that they may form pivots on which to hinge the defence of the rest of the position. They may be commanding features of the ground, groups of substantial buildings and enclosures, or wooded knolls giving cover from view and a good field of fire to front and flanks.

It is essential that such pivots should be capable of all-round defence, and that each should be able to sweep with fire a large proportion of the ground lying between it and those on either side.

If the pivots are strongly held and fortified, the number of troops necessary for the defence of the intervening ground need not be large, and the defence works themselves may be limited to ensuring that ground invisible to the defenders of the pivot groups is swept by fire.

A position, defended on these principles, lends itself to those local offensive movements which are so effective in keeping alive an offensive spirit in the defenders, in wearing down the attacker's powers, and in facilitating the delivery of a decisive counter attack.

4. The position, if extensive, should be divided into sections to each of which a name and number and a distinct unit or formation should be assigned (*see also* F.S. Regns., Pt. I, Sec. 108). Definite portions of such formations should be detailed as garrisons to each of the localities referred to above, in the defence of which every artifice of field fortification should be brought into play. It is in the preparation of these pivots that field engineers should generally be employed.

Within each section the commander may organize sub-sections, describing them in any manner best calculated to assist the troops. Example :—No. II (North East) Section, C sub-section (Green Farm).

5. The principal defences will, as a rule, consist of fire trenches which may be disposed in irregular lines or in groups with intervals, according to the character of the ground.

If the position consists of a succession of spurs and re-entrants, it may often be possible to draw back the trenches on the spurs, so that they are invisible from the foot of the spur on which they are placed. The fire from such trenches should be arranged so as to sweep obliquely the front of the spurs on either side, while a trench placed at the head of the valleys which divide them provides the necessary fire for searching the re-entrants. (Pl. 18.)

In tracing trenches which are to provide flanking fire, steps must be taken to protect them against enfilade fire, especially from artillery at long range.

6. To increase the volume of the defenders' fire, tiers of trenches may sometimes be employed ; while in other parts of the position trenches for companies or smaller units may be arranged in échelon which will make it difficult for the attackers' artillery to keep more than a small portion of the defenders under fire. There must, however, be no idea of using trenches, arranged in depth, as successive lines of defence. This must be clearly understood by the defenders of all trenches constituting the main zone of defence, which should be held to the last.

7. Careful arrangements are necessary to prevent the extreme flanks of an entrenched zone which do not rest on an impassable obstacle being quickly turned.

Such flanks should be gradually refused in échelon until some suitable supporting pivot is reached. They should not be turned back obliquely to the main front, as this exposes them to enfilade fire and does not seriously increase the circuit of a turning movement.

The allotment of a mobile force to the flanks in addition to such defences is a question which the superior commander must decide on the spot.

8. To facilitate inter-communication throughout the position, trenches, shelters, cross roads, etc., etc., should be provided with sign boards showing the nature of the place, the allotment of troops, etc. By night lanterns, extemporized from oblong wooden boxes, one side of which is covered with semi-transparent paper on which directions can be written, may be used for the same purpose. Arrangements should be made for visual signalling between adjacent fire trenches, and between the fire trenches and commanders of groups.

Telephones should be freely used when available.   (*See also* F.S.R., Pt. I, Sec. 86.)

9. With a view to misleading the attackers as to the situation and extent of the main position, and thus creating favourable opportunities for the defenders to manœuvre against them, an *advanced position* may be occupied. Such a position should be distinct from the main defensive position, and may consist of groups of trenches and artificially strengthened tactical points. Special instructions as to the amount of resistance to be offered, and as to the direction in which they are to retire, must be given to the troops occupying this advanced position.

10. A CLEAR DISTINCTION MUST BE DRAWN BETWEEN THE OCCUPATION OF AN ADVANCED POSITION AND OF ADVANCED POSTS. THE LATTER ARE LOCALITIES WHICH ARE OCCUPIED IN ORDER TO DENY GROUND TO THE ENEMY AND NOT MERELY AS A SCREEN TO THE MAIN POSITION. If they can be effectively supported by fire from the main position, they may be of value in breaking up an attack; but the fact that they form salients against which the enemy may be able to bring a heavy converging fire must be given due consideration when coming to a decision.   (*See* F.S. Regs., Pt. I, Sec. 108.)

11. The general duties of the commander of a group of posts in an entrenched zone are those laid down in F.S. Regns., Pt. I, Sec. 78, for a commander of outposts.

During the construction of the entrenchments he should study his surroundings so as to obtain information as to the strong and weak lines of fire, the points from which fire can best be controlled and communication with other posts arranged. Where practicable, he should examine the position from the attacker's point of view.

He should prepare a sketch of his command on as large a scale as possible, showing the general arrangements of the works, obstacles, kitchens, latrines, etc., the ranges to various range-marks, and the points from which adjacent works can best be supported.

This sketch, together with any other information such as work still to be carried out, special arrangements by night, and points where the enemy has been observed, must be handed over on each transfer of command.

51. USE OF VILLAGES IN DEFENCE.

(*See also* Military Engineering, Pt. I, Sec. 15.)

1. If it is decided to occupy a village, either in the main zone of a defensive position or as a supporting point in rear, every effort should be made to organize an obstinate defence.   Such places, strongly held, not only assist in breaking up the attack, but may be of great assistance in driving out the enemy, should he succeed in penetrating the position. It will, as a rule, be advisable to regard a village as a section of the position, and in no case should the main highway through a village form a boundary between adjacent sections of defence. The superior commander will decide whether a village, situated in a position, is to be occupied or not.

2. The suitability of a village for defence depends on :—
   (a) The nature of its surroundings.
   (b) The extent and shape of the village itself.

As regards the latter point, villages lying end-on to the enemy can often be made strong against flank attack, and may, therefore, be useful on the flanks of a position.

3. For the defence of a village, a definite garrison should be detailed under the command of a selected officer. The latter will be responsible for selecting the main and any interior lines of defence, for dividing the village into sub-sections, for allotting to each a proportion of the garrison, for arranging for a central hospital for wounded men, and for notifying the position of his headquarters. A general reserve should be retained in the hands of the commander to deliver counter attacks against any of the enemy's troops who may succeed in entering the village and to man the "keep" if one is prepared.

4. Each subordinate commander should consider the preparations for the defence of his sub-section in the following order :—
   (i) Improvement of the field of fire.
   (ii) Provision of cover and preparation of buildings for defence, much of which may be done concurrently with (i).
   (iii) Provision and improvement of communications.
   (iv) Provision of obstacles and barricades.
   (v) Arrangements for extinguishing fires.
   (vi) Ammunition supply.
   (vii) Food and water.
   (viii) Removal of sick and wounded.
   (ix) Retrenchment.

5. The firing line should usually be entrenched in front of any buildings to prevent casualties from shells which burst against their walls, and arranged to bring a powerful volume of fire on the best lines of attack.

6. Guns and machine guns should not, as a rule, be located amongst buildings, as there is considerable risk of their being discovered in such a position, in which case they are likely to become the object of concentrated hostile fire. Concealed positions in rear and to a flank should preferably be selected from which to flank a village and bring a cross fire on the enemy's probable lines of approach, and from which they can be more easily moved in accordance with the progress of the battle.

7. If, as is often the case, important bridges are located within the village definite instructions as to their value to the commander should be obtained before defence preparations are commenced

52. Use of Buildings in Defence.

(*See also* Military Engineering, Pt. I, Sec. 14.)

1. Buildings, when not exposed to artillery fire, may be of great defensive value. They should not be held when actually under artillery fire as they afford a conspicuous mark, and the danger from flying debris caused by shells is considerable ; but, if time and labour are available,

they should be prepared for defence, and occupied when the artillery fire ceases.

2. The principles laid down in Sections 1 and 27 are applicable to the defence of buildings, but the following additional points also require attention :—

   (i) Bullet-proof barricades to doors and windows. The means of exit, not necessarily on the ground floor, must be dealt with in a special way. It is easier to make loopholes in the barricades (Pl. 23) rather than to attempt to loophole the walls.

   (ii) Arrangements for ventilation, for the storage of ammunition, provisions and water, for a hospital and for latrines.

   (iii) Arrangements for extinguishing fires.

   (iv) Destruction of any outlying buildings which are not to be occupied, bearing in mind the importance of leaving no adjacent cover where an enemy might collect for assault.

3. If the building is large and strongly built, and it is intended to make an obstinate defence, arrangements must be made for interior defence by loopholing partition walls and upper floors made bullet-proof and strengthened if necessary to sustain the extra weight. Material with which to improvise additional cover or movable barricades to cover the retreat from one part of the building to the other, or from one building to another, must also be provided.

53. Use of Woods in Defence.

(*See also* Military Engineering, Pt. I, Sec. 16.)

1. Woods vary so much in character that it is impossible to give instructions for their use in defence which shall be suitable in all cases.

The two attributes common to most woods are the obstruction they offer to the passage of troops, whether in defence or attack, and the concealment they provide. As to the obstacle, it is the defenders' business to arrange that it shall cause the least inconvenience to his own, and the greatest inconvenience to the enemy's troops. The concealment afforded should be so utilized as to be almost entirely in favour of the defence.

Special precautions are necessary for the defence of woods which run down from a position towards the enemy, since they make co-operation between the artillery and infantry of the defence almost impossible and afford the enemy a covered line of approach.

2. In the case of large woods the difficulty of controlling troops engaged in them, the liability of those troops to lose their way, and the rapid consumption of reserves, make the improvement of their communications one of the first considerations.

If time is limited it will generally be best to devote it to improving the communications rather than to multiplying obstacles. For infantry several parallel tracks cleared of undergrowth are of more value and take less time to make than one broad road.

3. The front edge of a wood often has a boundary capable of being quickly made into a good fire position, but usually offers a good mark for artillery fire ; for this reason it may be desirable to place the firing line some 200 yards in advance, this being about the maximum distance

short of the wood at which shrapnel should be burst, in order to be effective.

If by clearing the undergrowth a good field of fire can be obtained between the tree trunks, the firing line may sometimes be placed with advantage 25 to 50 yards within the wood. In order to economize troops, those portions of the front from which effective fire can be developed should alone be occupied by the firing line, the remainder being entangled.

4. Where roads, rides or clearings exist in a wood the rear edge may be organized as a second line of defence. Such positions have often proved sufficient to stop the advance of troops who have succeeded in penetrating the front.

On the other hand, entrenchments and clearances on a large scale in the interior of a wood will seldom repay the time and labour necessary for their construction.

5. If defences in rear of a wood are more convenient than in front, the best arrangement will be to straighten and entangle the flanks and rear edge and take up an enfilading position some distance behind. The rear edge may be cut so as to leave well defined salients into which the attackers may crowd and so provide a good target. In making clearances, much can be done by judicious thinning, and large trees should not be felled. Communications throughout the wood should be blocked.

54. DEFENCE OF OUTPOSTS.

(See also F. S. Regs., Pt. I, Sec. 76.)

1. Outposts will always be entrenched, and advantage should be taken of natural cover to save time and labour in the construction of their defences. These works should, if possible, have an all-round field of fire, and protection against reverse fire, and should be surrounded by an obstacle. Where the outposts are within the field of fire of the main body, they must be protected from its fire in order not to mask it.

2. Pl. 30, Fig. 1, shows the section of a post for a piquet, which gives an all-round field of fire and overhead cover. It will rarely be possible to construct such elaborate works, but the principle of construction will be the same for works of the most elementary order.

3. Pl. 30, Fig. 2, shows the plan of a post for four men which does not require a high command and from which fire can be delivered in any direction.

55. LINE OF COMMUNICATION DEFENCES.

(See also F.S. Regs., Pt. II, Sec. 11.)

1. Campaigns often entail a long line of communications in a more or less hostile country. Even when protected by a field army this line is liable to raids and must, therefore, be protected by fortified posts. These posts may cover (a) a comparatively large area of ground in order to afford protection to supply depôts, convoys and transport animals, or, in the case of a railway, to rolling stock, station buildings,

telegraph stations, &c.; (b) a very limited area in order to defend a bridge, a signalling station, or similar object

2. AS EVERY MAN EMPLOYED ON COMMUNICATIONS MEANS ONE LESS IN THE FIELD ARMY, THE GARRISONS OF SUCH POSTS MUST BE KEPT AS LOW AS POSSIBLE, AND EVERY EFFORT MUST BE MADE TO ECONO-MIZE MEN BY THE SKILFUL USE OF GROUND AND FIELD FORTIFICATIONS. Sometimes the site of a post may have to be in a spot chosen for reasons other than tactical ones; and careful dispositions, including the provision of good artificial protection, will then be more than ever necessary.

8. A proportion of the garrison should, if numbers permit, be kept in hand and only as a last resource should the defensive perimeter be manned by the whole force.

When organizing the defence of a post special attention should be paid to the following points:—

   (i) Defenders should be quartered close to the positions they have to man.

   (ii) Arrangements for storage of ammunition, water and supplies.

   (iii) Provision of strong obstacles.

   (iv) Adequate cover with a clear field of fire.

   (v) Provision of automatic alarms, if possible.

   (vi) Good communications, including telephones, telegraphs, and signalling.

5. Plenty of time is usually available for the organization of the defence, and, given adequate supplies of ammunition, food, water and material, small posts can be made practically impregnable against raid attacks, even when supported by artillery; while larger posts can be occupied in such a manner that the risk and loss involved in an attempt to force a way into them, even under cover of darkness, would rarely justify the attempt.

6. The garrisons of isolated works cannot, as a rule, be reinforced from a distance when attacked at night, and for this reason it is essential that they should be self-contained.

7. The main defences of a post of class (a) will consist of a ring of closed and self-contained works supporting each other. The number of these works and their distance from the centre will depend on the ground and the garrison available. The intervals should, if possible, be closed by a strong obstacle, flanked by fire from the works.

In most cases an inner line of defence will be required, and possibly a "keep." This inner line will not be so elaborate as the outer, and will generally consist of fortified houses, garden enclosures, small block-houses, &c., placed in the outskirts of the village or depôt and arranged so as to sweep all approaches and internal communications. The strongest work in the post will form the "keep" irrespective of its position in the inner or outer lines of defence; and reserves of supplies, ammunition and water should be arranged accordingly.

8. The posts of class (b) will consist of one or more closed works, whose garrisons may vary from 6 to 50 men.

9. The types of works will necessarily depend on the nature of the probable attack as well as on the materials available. Invisibility is not essential. Head cover is necessary, and overhead cover often

desirable; and, since the attack may come from any direction, protection against enfilade and reverse fire must be provided. If the enemy is provided with artillery, well-concealed deep trenches and splinter-proof cover must be provided (unless the ground affords adequate cover close at hand). Against rifles only, bullet-proof blockhouses will suffice. (Pl. 26.) Against badly armed savages, stockaded enclosures may be enough. (Pl. 24.)

Pl. 27 gives an example of a defensible post where the blockhouses are arranged to enfilade the lines of obstacles. South Africa produced corrugated iron and shingle blockhouses surrounded by barbed wire (Pl. 26); on the north-west frontier of India, stone sangars are the rule; in the Lushai Expedition of 1889, bamboo stockades were employed (Pl. 24); in the Soudan, breast-works of sand and thorn zerebas. Where railway stations have to be protected, blockhouses, stockades and splinter proofs made of rails, and loopholed buildings will predominate. (Pl. 25.)

In savage warfare hints as to the best design of defensive work may generally be got from the enemy, who will have evolved the types best suited to local materials, as well as to resist the form of attack and weapons which he will employ against us. Such types, when improved by the light of our own knowledge, modified to suit our weapons, and executed with the aid of good tools and engineering skill will, as a rule, be suitable for our own use.

10. Stockades are improvised defensible walls, which, in addition to affording cover to their defenders, form a fair obstacle to assault. They are only suitable for passive defence in positions where they are not exposed to artillery fire. The commonest forms of stockade consist of earth, gravel or broken stones, &c., between two walls of planking, corrugated iron, &c. The necessary thickness can be obtained from the table in Chap. I, Sect. 2. Types of stockades of rails and sleepers are shown on Pl. 25.

11. The loopholes through which the defenders deliver their fire should be so arranged that the enemy, if he succeed in surmounting any obstacle that may have been made, will not be able to use them from the outside.

12. Automatic alarms and flare lights are useful against night attacks, (see Section 42 (4)). They should be used in combination with obstacles, if any, and either protected or concealed, so as to prevent the enemy removing them. Flare lights should be screened in rear so that the defenders may remain in shadow.

One of the simplest alarms is a row of tin pots, each containing one or two pebbles, hung on a wire fence so as to rattle when the latter is disturbed. Holes may be dug in the ground in which to hang these pots, in order to render them invisible to the attack.

Pieces of tin, such as the tops of tin pots, bent round the wires of an entanglement, answer the same purpose. In rain or wind, however, these devices make such a continual noise that they finally cease to attract attention and also prevent the sentries using their natural powers of hearing.

Pl. 28 shows how a rifle alarm can be fixed in connection with a wire fence entanglement or trip wire.

A spring-bell (such as a bicycle bell) usually forms a better alarm signal than a rifle cartridge in places where the discharge of a rifle is not an unusual occurrence.

A trip wire, or one of the wires in a fence, can be arranged as in Pl. 29 to fire a cartridge, which in its turn will ignite a flare. To connect the flare with the cartridge, one end of a piece of instantaneous fuze is inserted into an empty cartridge case and pushed well up to the cap, the mouth of the case being pinched to keep the fuze in position. The other end of the fuze is bound up with loose cordite or flaked gun-cotton, a little dry tow arranged round it, and the whole covered with tow, straw, dry grass, etc., saturated with oil.

Alarm signals and flares should always be so arranged that they can be operated at will by the defence, and not only by the enemy.

13. On a dark night, it is difficult to ensure the men's rifles being aimed in the required direction. Any device to assist them in this matter is useful. Fixed rifle rests may be made, or failing these, some such device as a wooden bar can be arranged across the loopholes, to prevent a man raising his rifle barrel too high. Posts painted white on the defenders' side make a good aiming mark, if the night is not too dark.

14. Particular attention should be paid to the entrances of closed works. They may be closed by a gate, barbed wire or other movable obstacle. When wire is used, a good plan is to construct a winding approach, making access by night difficult. (Pl. 26, Fig 3.) In all cases entrances must be covered by the fire of the defence. Entrances to admit vehicles require a width of 8 feet, and may be closed by chevaux de frise.

56. Defence of Camps.

(See also F. S. Regs., Pt. I, Secs. 143 and 154, and Military Engineering, Pt. I, Sec. 17.)

1. When operating in mountainous country against an uncivilized enemy who is likely to make night attacks, the three principal points to attend to are: (*a*) an outer line of strong self-contained piquets, placed so as to deny to the enemy all ground from which he could fire into the camp; (*b*) a well-defined firing line, round the camp itself; (*c*) a good obstacle in connection with it.

2. When the number of the defenders is insufficient to provide an all-round defence, the perimeter must be defended by flanking fire from works constructed with that object, and especial attention must then be paid to providing as formidable an obstacle as possible.

3. The positions to be taken up in order to repel a night attack should be marked out as soon as possible after the force has reached camp. If there is only time to do this with a line of stones, it will give the defenders a definite line to occupy and hold on to.

4. For convenience in camping, troops should generally occupy the same relative positions each night; but this convenience must be sacrificed to the arrangements necessary for defence, as it is very important that units should camp close to the ground which they would have to hold in case of attack.

5. In selecting a camp site attention must be paid to the water supply and its protection; but the first consideration is a good position against possible night attacks.

6. When operating in open country not of a mountainous nature, or in bush warfare piquets will often be withdrawn by night. In such cases, a well defined firing line combined with an efficient obstacle must be provided round the camp. It may, under certain circumstances, be advisable to apply this principle whenever the force halts in order to provide an asylum for the non-combatants and transport. These improvised defences may take the form of Laagers or Zeribas. Laagers are enclosures formed with the vehicles accompanying a force, supplemented by breastworks of pack saddles, stores, etc., and strengthened with trenches and abatis. Zeribas are enclosures fenced in by abatis of thorn bushes. It being most important to obtain a clear field of fire, the bush nearest the site of the camp must first be removed and arranged round the perimeter. Subsequently tracks and hollows, by which the enemy might approach, may be filled with thorn scrub if time allows.

7. In all cases a cross-fire should be arranged to sweep the lines of obstacle and the most likely avenues of approach; for this purpose machine guns are especially valuable. The general trace of a laager or zeriba should, as a rule, be in accordance with the principles shown in Pl. 27.

CHAPTER VIII.—FIELD LEVEL AND FIELD GEOMETRY.

(*See also* Military Engineering, Pt. I, Sec. 8.)

57. FIELD GEOMETRY.

1. In some of the more technical operations of field engineering, such as the construction of bridges and in road and camp work, a knowledge of the following applications of simple geometry in the field will often be of use. No special instruments are required for this purpose.

2. To lay out a right angle. Let X be a point in a given straight line A B (Pl. 81, Fig. 8), from which it is required to set off a right angle.

From X measure off a distance of 4 units X C along A B. Take a piece of line or tape 8 units long, apply one end to point X, and the other to point C; find a point in the tape 8 units from X, and, seizing it at this point, draw the bight out to D, till the line is taut, then C X D is a right angle.

8. To trace a perpendicular to a given line from a point outside. Let X be the point outside the line A B (Fig. 4), from which it is required to draw a perpendicular to that line. Take a tape or cord longer than the perpendicular will be; fix one end at X, and stretch it taut, so that the other end shall cut A B in C. Drive in a peg at C, find D, the middle point of C X. With D as centre, swing D X or D C round to the position D E, cutting A B in E. Join X E, then X E is at right angles to A B.

4. To lay off an angle of 60° or 120°. Let X be the point in the line A B (Fig. 5) from which it is required to lay off an angle of 60°. Take any point C in A B towards that end of the line from which the angle of 60° is to be drawn. Take a tape or cord twice the length of X C, and fasten the ends to X and C. Seize it at the middle and draw the bight out taut to E. Then the angle E X C is 60° and A X E is 120°.

5. To bisect a given angle. Let A B C be the angle which it is required to bisect (Fig. 6). In A B take any point D. Fasten the end of the tape at D, and take it round B and back again to D. With the length thus found mark E in B C and make the loose end fast at E. Take the centre of the tape from B and stretch it tight in the position D F E. B F will bisect the angle A B C.

6. To lay out an angle equal to a given angle. Let X be the point in the straight line A B (Fig. 7), from which it is desired to lay off an angle equal to the angle D E C. In the bounding lines of the angle D E C take any two points D C, and from X measure XG equal to EC. Take a tape equal to C D E. Put the ends at C and E, and make the tape cover C D E. Holding the tape by the point above D, transfer the ends which were at E and C to X and G respectively, and pull the tape taut. Then the point which had been at D will be at some point F, and the angle F X G will equal the angle D E C.

7. To find the distance between any two points A and B when it cannot be measured directly. From B (Fig. 8) lay off the line B D as nearly at right angles to A B as possible, D being at any convenient distance. In B D select a point C so that B C is some multiple of C D. From D lay off the angle B D F equal to the angle A B D, and on the opposite side of the line B D. Make D E of such a length that the point E is in line with A and C.

Then A B : B C :: D E : C D,

or $A B = \dfrac{B C \times D E}{C D}$

58. FIELD LEVEL.

1. In various operations of military engineering a special instrument called the field level is used for setting off angles on the ground and for gauging slopes. It is shown on Pl. 32, and can be used (1) when closed, as an ordinary spirit level for levelling, etc., the spirit level being on the edge of the limb *c* (Fig. 2) ; (2) when open, as a square for setting off right angles (Fig. 1) ; as a protractor for setting off angles ; as a mason's level with plumb bob ; or for setting off a slope of any gradient (Fig. 3). It is especially useful as a level by which the depth of various points in a hollow or bed of a stream, &c., can be read by sighting the level on a marked pole held at the various points in turn. (Fig. 4.)

CHAPTER IX.—CAMPING ARRANGEMENTS.

59. FIELD KITCHENS.

(*See also* Military Engineering, Pt. V, Sec. 1.)

1. The simplest arrangement for cooking in the field for any party over 20, if the halts are not of long duration, is to place a proportion of the kettles on the ground in two parallel rows about 9 inches apart, handles outwards, block the leeward end of the trench so formed with another kettle, lay the fire between the kettles and place one or two rows of kettles on those already in position. (Pl. 33.)

Mess tins can be arranged similarly, but in their case not more than eight should be used together.

2. The most economical method where time is available is to dig or raise a narrow sloping trench for the fire on which the kettles are placed. The interstices are then filled up with stones and clay so that the fire, fed from the windward end, may draw right through. A chimney may be built at the other end to increase the draught. (Pl. 34, Figs. 1 to 4.)

The chimney can be built of sods, supported where it passes over the trench by flat stones, slates, wood covered with clay, &c. The insides of the trench and of the chimney may be plastered with clay to make them last longer. Several such trenches may be combined, as shown in Fig. 2, to form what is known as the *parallel* or *rectangular* kitchen, or three trenches may converge to one flue, as shown in Fig. 3, forming what is known as the *broad arrow* kitchen.

3. Figs. 5 and 6 give details of a covered kitchen, suitable for standing camps. The roof may be covered with tarpaulins, or in the manner described in Sections 64 and 65.

60. FIELD OVENS.

1. The simplest form of a field oven consists of a hole burrowed in firm ground in the face of a bank (Pl. 35, Fig. 5), or the hearth may be sunk below the ground surface, with an arch formed by a hurdle or sheet iron on the top of which earth is piled (Pl. 35, Figs. 1 to 4). As sheet iron is liable to become distorted under the weight of the covering and heat of the fire, it is advisable to stiffen it by ribs of iron, if procurable. An old cart wheel tyre cut in two is very useful for this purpose. The two gable ends can be formed with sods. When a hurdle is used it should be well plastered on the outside with a mixture of cow dung and clay so as to leave an arch when it burns away. The entrance to the oven is closed either by a hurdle plastered with clay or by sods.

This oven is specially suitable for making bread, and will bake for about 150 men at a time.

61. Water Supply.

(*See also* F.S. Regs., Pt. I, Sec. 57 and Military Engineering, Pt. I, Secs. 3 to 7.)

For stores, *see* Appendix III, Table 13.

1. A daily average of one gallon per man is sufficient for drinking and cooking purposes.

In standing camps, an average allowance of 5 gallons should be given for a man, and 10 gallons for a horse.

A horse, bullock, or mule drinks about 1½ gallons at a time.

In making calculations of the time and space required for watering horses, each animal should be allowed five minutes at the trough, and four feet of lateral space.

The service 33 ft. trough holding 600 galls. will water 16 horses at a time, using both sides of the trough.

2. The rough average yield of a stream may be measured as follows :— Select some 15 yards of the stream where the channel is fairly uniform, and there are no eddies. Take the breadth and average depth in feet in three or four places. Drop in a chip of wood and find the time it takes to travel, say, 30 feet: Thus obtain the surface velocity in feet per second. Four-fifths of this will give the mean velocity, and this multiplied by the average sectional area in square feet will give the yield per second in cubic feet of water (one cubic foot equals six and a quarter gallons).

3. The source of the water supply should be carefully investigated, and measures taken to prevent the pollution of the water *en route* to the drinking supply. Water from small ponds and shallow wells should be avoided.

4. Water near the surface should be sought for in hollows where the earth is moist or the grass unusually green, or where the thickest mists rise in the mornings or evenings; in valleys just above a contraction in width or at the point of junction with a branch valley. Where a line of rocks just shows above ground, water will often be found by digging on the uphill side of the rocks and not too close to the outcrop.

5. If the supply be from springs, each springhead should be opened up and surrounded by a low puddled wall to keep out surface water. Casks or cylinders made of brushwood, like gabions, make good linings for springs. After they have been placed in position, puddled clay may be worked down between them and the banks. The overflow may be received into a succession of casks or half barrels (which may with advantage have their insides charred) let into the ground close together, the overflow from the first passing into the second, and so on ; or deep narrow tanks with puddled sides may be constructed to catch the overflow.

6. If the supply be from a lake, pond, or stream, separate watering-places for men and animals must be marked out and sentries posted.

In a stream the men should draw drinking water above the place for the animals; while washing, &c., should be done below it. Barrels sunk in the bed of a small stream afford convenient dipping places. Water bottles, camp kettles, and similar articles liable to convey dirt must on no account be dipped direct into drinking water tanks, &c. Clean tins or dippers must be provided if draw-off taps are not available. Drainage should be disposed of as far below the sources of supply as possible.

7. The "lift and force pump," weighing 84 lbs. complete, is the pattern in general use in the service. It is worked by two men. It can lift water from a maximum depth of 28 feet, and force it 32 feet (*i.e.* 60 feet in all), at a rate of 12 gallons a minute. Two are carried in the field by each Field Troop, and four by each Field Company of Engineers. To obtain the best results the height of lift or suction should be reduced to a minimum, and can rarely exceed 20 feet. The end of the suction pipe must never be allowed to rest on the silt or mud at the bottom of a well or stream.

62. Arrangements for Watering Animals.

1. All the horses in a camp should be watered in an hour. The sites should be arranged so that the animals may move to and from them without confusion or crowding, arriving from one direction and leaving in another.

2. Mules, being particularly fastidious about drinking water, should, if possible, be allotted a separate drinking place. Stagnant water, as in a pond, is apt to be contaminated by large numbers of animals going in to drink; and even in running water, when many animals are drinking, those down stream get foul water. If possible, therefore, the water should be run into drinking troughs made of canvas or boards, but trenches lined with puddled clay will also answer the purpose.

The overflow from the troughs must be carried off with the surface drainage, and the ground on both sides of the troughs should, if possible, be metalled or covered with brushwood and drained for a width of 10 feet.

3. When troughs cannot be made, the banks should be cut down, and a hard surface formed on the ramp to prevent it being cut up by the animals, and a barrier may be put up to prevent them from going too far into the water. The water must not be less than 5 or 6 inches deep.

63. Methods of Purifying Water.

1. The best method of purifying water is by boiling, which gets rid of temporary hardness, renders dissolved organic matter harmless, and, when properly carried out, practically destroys all micro-organisms. The water should be kept at the boil for at least five minutes. Boiled water should be aërated before drinking. This can be done by passing it through a sieve. Empty biscuit tins pierced with small holes suspended over a storage tank do very well for this purpose, but care is necessary to prevent the addition of fresh impurities during aeration and distribution.

2. As it is not always possible to provide means of boiling water on a large scale, resort must be had to filtration. Formerly mechanical filtration only was attempted and a clear sparkling water was considered good. Efforts are now directed to removing chemical as well as other impurities.

3. The Slack and Brownlow is the pattern of filter in general use. It consists of three main parts, viz., the sterilizing filter, the clarifying filter, and the pump. (*See also* App. A, Military Engineering, Pt. V.)

Treated with care these filters work satisfactorily, and will deliver up to 34 pints in 10 minutes, so long as the filter is clean. They are carried in the field by field ambulances at the rate of 2 per cavalry ambulance, and 6 per field ambulance. These filters form the filtering medium of the army pattern filter water cart.

The service water cart (M. IIA or IIB) contains 114 gallons. They can be filled by their own pumps in 20 and 12 minutes respectively, and deliver 8 or 4½ gallons of filtered water per minute by means of 12 taps.

4. Dirty water should be strained before filtering. A good method is to tack a sheet on to a wooden frame so as to form a bag or basin in the bottom of which are put a couple of handfuls of wood ashes. The water is then poured on to them and allowed to percolate into a receptacle beneath.

5. Chemicals are sometimes added to precipitate suspended matter, to remove hardness, or to oxidize organic impurities.

Muddy water may be cleared by adding alum. Six grains of crystallized alum per gallon or one teaspoonful to ten gallons is sufficient. It should be added some hours before the water is required.

Water can be softened by the addition of limewater for drinking, and carbonate of soda for washing purposes. The latter is unsuitable for drinking water, as it gives an unpleasant taste.

Permanganate of potash (Condy's fluid) removes offensive smell from water and to some extent oxidizes dissolved organic matter. It should be added until a faint tint remains permanent. It has not a disagreeable taste.

64. SHELTERS.

(*See also* F.S. Regs., Pt. I, Secs. 55 and 120, and Military Engineering, Pt. V, Sec. 2.)

1. Simple shelters may be formed in many ways. One method is to drive two forked sticks into the ground with a pole resting on them; branches are then laid resting on the pole, thick end uppermost, at an angle of about 45°, and the screen made good with smaller branches, ferns, &c. A hurdle may be supported and treated in a similar way.

2. A shelter tent for four men may be formed with two blankets or waterproof sheets laced together at the ridge, the remaining two blankets being available for cover inside.

3. When no other materials than earth and brushwood are available, a comfortable bivouac for 12 men can be formed by excavating a circle with a diameter of 18 feet, or thereabouts, and building up the earth to form a wall 2 or 3 feet high. The men lie down, like the spokes of a

wheel, with their feet towards the centre.   Branches of trees, or brushwood stuck into the wall, improve the shelter.

65. BRUSHWOOD HUTS.

1. If a camp or bivouac is likely to be occupied for some time, the
men ought to be hutted. In the tropics huts should be raised off the
ground level.

2. The most convenient form of hut is rectangular in plan, with
sufficient width for two rows of beds and a passage down the centre,
but, where the material available is of small size, one row of beds may
be provided, or the hut may be made of circular form. A width of at
least 6 feet should be allowed for each row of beds, and the passage
should be from 2 feet to 4 feet wide.

The accommodation may be calculated on active service at one man
per foot of length when there are two rows of beds, and one man to every
2 feet when there is only one row.

3. When brushwood of 2 or 3 inches diameter and 14 or 15 feet long
is available, a hut for a double row of beds may be made as in Pl. 36.

The size of the hut having been settled, a section is laid out on the
ground, and from this the length of the rafters is obtained. Each side
of the roof is then made separately as follows (Pl. 37) :—

Poles of 2 inches diameter are laid on the ground parallel to each
other, from 18 inches to 2 feet apart, as *aa* in Fig. 4.   These
form the rafters. On the rafters, and at right angles to them, light
rods or purlins, *bb*, from ½ inch to ¾ inch thick are laid, the uppermost
one being at such a distance from the end of the poles as will allow the
frames to lock at the desired height above the ground, the lowermost
one being within a few inches of the bottom. Between the top and
bottom purlin others are placed at a distance apart equal to slightly less
than half the length of the thatching material, so that the latter may
be supported at three points. With good wheaten straw the interval
may be from 12 inches to 1½ feet.   At each point of crossing the purlins
and rafters are secured by a short length of spun yarn, and the frame
thus made is afterwards stiffened by diagonals lashed underneath the
rafters.

The roofing material, which may be unbroken straw, rushes, long
ferns, &c., is now put on.   Commencing at the bottom, a layer 4 or
5 inches thick is laid evenly over the three lowest purlins, ears or tops
downwards, and secured by a light rod or *thatching piece* tied with
spun yarn at intervals of 2 or 3 feet to the second purlin from the
bottom. A second layer is now put on one purlin higher up, and is
secured in a similar way to the third purlin from the bottom, and so on
until the top is reached ; the last layer should project over the top purlin,
so that when the frames are locked the ends may be twisted together
to keep out the wet.   When both frames are ready they are raised and
locked, as in Fig. 3. Forked uprights and a ridge piece may be added to
stiffen the roof.

Each side of the roof may be made in one or more sections as
convenient. The ends of the purlins should project about 2 feet beyond
the outside rafters, and are supported by the framework forming the

gable ends, Fig. 5. The latter are made and thatched in a similar way to the roof, and simultaneously with it. Openings should be left for a door, and close under the ridge at the gable ends for ventilation.

4. Huts may also be thatched by forming the straw or grass into *panels*. The straw in moderately thick layers is doubled and nipped near the centre between two rods, which are tied together tightly at the ends and at intervals of about 6 inches.

The panels thus formed are tied on to the roof, being placed so as to overlap like large slates.

5. In order to give additional head room, the passage may be sunk as in Pl. 37, Fig. 3, with steps at each end, the earth being used to raise the roof. In very cold weather the whole interior of the hut may be excavated, fireplaces constructed as in Pl. 36, Fig. 2, and, if the rafters be strong, some of the excavated earth may be thrown on to the roof, a series of collar ties being added to strengthen it.

6. Walls may be constructed of *wattle and daub, i.e.*, continuous hurdle work daubed over on one or both sides with clay, in which is a proportion of any fibrous substance, such as straw, grass, horse hair, &c., chopped into short lengths to prevent the clay cracking and opening as it dries. This mixture, which should be kneaded into the consistency of a stiff paste, should be worked in with the hands. The sides should be strutted at intervals to resist wind, and the roof may be carried on a ridge pole, which may be strengthened by uprights in the centre. (Pl. 36, Figs. 3 and 4.)

66. Log Huts.

1. When straight timber is abundant, log huts may be constructed. The walls are formed of logs laid horizontally. They are notched into each other at the corners, each notch extending half way through the log, so as to leave every layer flush. No fastenings are required beyond some *trenails* (wooden pegs) to secure the rafters to the top logs. The roof may be made as already described, or the covering material may be of slabs of wood, or bark of such trees as birch, &c.

2. Bark may be got off trees in large strips by cutting round the tree with a knife at intervals of about 4 feet. The width of bark required is then cut and beaten with a flat piece of wood to detach it from the tree.

67. Latrines.

(*See also* F. S. Regs., Pt. I, Secs. 56 and 60.)

1. Latrines should be dug as soon as possible after the troops reach their camp or bivouac. Latrine trenches should be arranged in one line as shown on Pl. 38, with 2½ ft. clear space between each trench. The size of each trench should be 3 ft. long, 1 ft. broad, and 1 ft. deep.

After use on the second day these trenches are filled in, and fresh ones dug in the intervals. On the third day a fresh row similar to the first is commenced 1 foot to the front and parallel to it.

If the ground available is limited in extent, the trenches may be increased to 2 ft. in depth, and made to last over the second day.

2. In excavating the trenches, the turf covering each should be removed and placed about 3 ft. behind each trench, the sides should be kept vertical, and the excavated earth should be well broken up and piled up close to the trench.

3. As a rule 5 trenches should be provided for 100 men for one day, but 15 trenches (*i.e.*, 3%) will suffice for a strength of 500 men.

If material is available, the trenches should be surrounded by a screen with an overlapping entrance formed in the centre of one of the sides. The length of screen necessary to surround latrines for 1,000 men calculated on a 5% basis, is about 130 yards.

4. When for exceptional reasons, deep trenches are necessary, they should be provided with pole seats, which should be cleaned daily, preferably with cresol solution. These latrines should be constructed to seat 5% of the troops, 1 yard per man being allowed. (Pl. 20, Fig. 3.)

5. It is very important that a couple of inches of the driest earth obtainable should be thrown each day into latrine trenches in use. If means are available, the earth may be artificially dried for this purpose. If carefully carried out, this will obviate all smell, and will tend to prevent flies collecting. The use of kerosene oil or lime, in and around the trenches, will still further assist in keeping flies away.

Small shovels or improvised scoops should be provided, when possible, in the proportion of one for every two short trenches.

6 On vacating a camp or bivouac, the position of latrines should be marked with the letter L formed with stones, &c.

68. Refuse Destructors.

All camp refuse should be collected and burned. Tins should afterwards be buried. Pl. 38, Fig. 2, shows a type of refuse destructor.

With this type and a 40 gallon tank provided with a tap, water can be boiled for drinking purposes and earth dried alongside for use in the latrines. When continual use is to be made of refuse destructors they should be lined with clay and placed under cover from rain.

CHAPTER X.—KNOTTING AND LASHINGS.

(*See also* Military Engineering, Pt. III.)

69. KNOTS.

1. The following are the most useful knots and their principal uses:—
   - (a) To make a stop on a rope, or to prevent the end from unfraying, or to prevent its slipping through a block; the *thumb knot* (Pl. 39, Fig. 1) or the *figure of* 8 (Fig. 2).
   - (b) To *bend* or join two ropes together. The *reef knot* (Fig. 3) for dry ropes of the same size; the *single sheet bend* (Fig. 4) for dry ropes of different sizes; the *double sheet bend* (Fig. 5) for great security or for wet ropes of different sizes, and the *hawser bend* (Fig. 6) for joining large cables.
   - (c) To form a loop or *bight* on a rope which will not slip. The *bowline* (Figs. 7 and 8) for a loop at the end of a rope, the *bowline on a bight* (Fig. 9) for a loop in the middle, with a double of the rope. The loop formed by passing the running end through a bowline loop at the end of a rope is called a *running bowline*.
   - (d) To secure the ends of ropes to spars, pickets, &c., or to other ropes (Pl. 40).

     Clove hitch (Figs. 1 and 2) (two half hitches) generally used for the commencement and finish of lashings.

     Timber Hitch (Fig. 3) for holding timber, &c., where the weight will keep the hitch taut.

     Two half hitches (Fig. 4) for making fast the running end of a rope to its standing part.

     Round turn and two half hitches (*or rolling bend*) (Fig. 5) for *belaying* (or making fast) a rope so that the strain on the rope shall not jam the hitches. This will be used for making fast a rope to a bollard or anchorage. Should the running end be inconveniently long, a bight of it should be used to form the half-hitches.

     Fishermen's bend (Fig. 6), for making fast when there is a give-and-take motion, *e.g.*, for bending a cable to an anchor (Pl. 56. Fig. 1).
   - (e) To fix a spar or stick across a rope (Pl. 40).

     Lever hitch (Fig. 7), for drawing pickets by a lever and fulcrum, fixing the rounds of a rope ladder, fixing bars to dragropes, &c,
   - (f) To form a loop on a dragrope.

     Man harness hitch (Figs. 7 and 8), the loop being of a size to pass over a man's shoulder.
   - (g) To secure a headrope, boat's painter, &c., to a post, ring, or rope, so that it can be instantly released (Pl. 41, Figs. 1, 2 and 3). *Draw hitch*, made as follows:—Pass a bight of the running end, (a Fig. 1), round the holdfast; pass a bight of the

standing part, (b Fig. 2), through bight a, and haul taut on
the running end; pass a bight of the running end, (c Fig. 3),
through bight b and haul taut on the standing part. This knot
will stand a give-and-take motion and can be instantly
released by a jerk on the running end.

(h) To fix a rope with a weight on it rapidly to a block.

Catspaw in the middle of a rope (Fig. 5), for hooking on a block.

(k) To transfer the strain on one rope to another.

Stopper hitch (Fig. 4), for use on occasions when it is necessary
to shift the strain off a rope temporarily.

70. SLINGS.

1. To sling a cask horizontally.   Make a long bight with a bowline
and apply as shown in Pl. 41, Fig. 6.

2. To sling a cask vertically (Fig. 7).   Place the cask in a bight at the
end of the rope, and with the running end make a thumb knot round
the standing part of the rope.   Open out the thumb knot and slip
it down the sides of the cask.   Secure with a bowline.

71. LASHINGS.

1. A rack lashing consists of a length of 1½-inch rope, with a pointed
stick at one end.   It is used for fastening down ribands at the edge of
the roadway of bridges.   It is commenced with a thumb knot at a
(Pl. 42, Fig. 1), the end twisted in the bight.   The stick is then put into
the bight, twisted against the hands of the clock till all is taut (Fig. 2),
and finally jammed in from right to left between the lashing and the
outside of the riband (Pl. 51, Fig. 1).

2. To lash one spar square across another, commence by a clove hitch
on spar a below b (Pl. 42, Fig. 4) and twist the ends together, carry at
least four times round the spars, as shown in figure, keeping outside
previous turns on one spar and inside on the other; two or more
frapping or cross turns are then taken, the corners of the lashings
being well "beaten in" during the process, and finished off with two
half hitches round the most convenient spar.

3. When the spars are the leg and transom of a trestle or frame, the
clove hitches should be on the leg below the transom and the lashings
should be finished off on the transom outside the leg.   When the
spars are leg and ledger the clove hitch should be on the leg above the
ledger.

To lash two spars together that tend to spring apart.   Begin with a
timber hitch or running bowline round both spars and draw them
together, then take three or four turns across each fork and finish with
frapping turns and two half hitches (Fig. 5).   When the spars are not
horizontal, the lashing should be finished off above the junction.

4. Wedges with well rounded points are often useful for tightening
lashings.   They are generally used by builders in scaffolding, and
should be driven in at the top of the lashings.

5. Hemp-rope lashings soon become loose, and require frequent
re-making.   Wire lashings should be used in their place when possible.

These can be made in a similar way to hemp-rope lashings ; but, unless staples are available, the wire should be finished off round a set of returns, and jammed between them and the timber. It is of little use attempting to finish off on a round spar. No. 14 gauge steel wire may be taken to have $\frac{1}{4}$ the strength of 2" cordage.

6. To lash a block to a spar.—The back of the hook is laid against the spar, a clove hitch is taken round the spar above the hook, then several turns round the hook and spar, and finished off with two half hitches round the spar below the hook. (Pl. 42, Fig. 6.)

7. The hook of a block is *moused* by taking some turns round it with spun yarn or very light lashing, commencing with a clove hitch on the back of the hook and finishing off with one or two frapping turns and a reef knot (Pl. 43, Fig. 1). (*See also* G.A.T., Vol. III, Sec. 4.)

8. The end of a rope is *seized* to the standing part with spun yarn or string, by forming a clove hitch round one of the ropes with the spun yarn near its centre, taking each part round both ropes in opposite directions, leaving one end long enough to take two frapping turns between the ropes, and connecting the two ends with a reef knot. (Pl. 41, Fig. 4.)

72. HOLDFASTS.

(*See also* Military Engineering, Part III, and G.A.T., Vol. III, Chap. 3.)

1. A picket used as a holdfast must be driven into the ground at right angles to the direction of the strain. If the latter is great, three or more pickets can be driven in a cluster to form a bollard.

2. If a large piece of timber is used as a bollard, its corners must be rounded off. Pl. 42, Fig. 7, shows a method of using a buried log for large strains. A trench having been dug long enough to hold the log, the cable is given one complete turn round it and passed up through a narrow incline constructed at right angles to the trench. The running end is then seized to the standing part in two or three places.

3. A 3. 2. 1. holdfast, made of 5 feet pickets, driven 2 feet 6 inches to 3 feet in the ground, should stand a strain of 2 tons. (Pl. 42, Fig. 3.)

73. STRENGTH AND SIZE OF CORDAGE, ETC.

1. The size of a rope is denoted by its circumference in inches, and its length is given in fathoms. (A fathom is 6 feet.) Cordage is usually issued in coils of 113 fathoms, and steel wire ropes in coils of 100 fathoms.

2. For field purposes, the safe working load for all cordage has been laid down as C^3 cwts., while for steel wire rope it may be taken as 9 C^2 cwts., where C is the circumference in inches. Steel wire rope may be taken as twice as strong as iron wire rope.

3. The strength of wire varies greatly; as a very rough rule it may be taken that the breaking weight in pounds equals three times the weight per mile in pounds. (*See* Appendix III, Table 14.) This rule holds good for iron and hard drawn copper wire, while steel wire may be taken as about twice as strong.

4. The strength of a lashing around two objects may be taken as $\frac{2}{3}$ of the number of times the lashing passes from one object to the other multiplied by the unit strength of the lashing, *e.g.*, a square lashing with four turns has a holding power of $\frac{2}{3} \times 16 \times$ strength of lashing ; in the case of a hook lashed to a spar with four turns it is $\frac{2}{3} \times 8 \times$ strength of lashing.

When using wire in lashings, multiply by $\frac{3}{4}$ instead of $\frac{2}{3}$.

CHAPTER XI.—BLOCKS, TACKLES AND USE OF SPARS.

(*See also* Military Engineering, Pt. III.)

74. BLOCKS.

(*See also* G.A.T., Vol. III, Sec. 60.)

1. Blocks are used for the purpose of changing the direction of ropes or of gaining power.

They are called single, double, treble, &c., according to the number of sheaves which they contain. The sheaves revolve on a pin, which should be kept well lubricated.

Snatch blocks (Pl. 43, Fig. 1) are single blocks with an opening in one side of the shell, to admit a rope without passing its end through. This opening is closed by a hinged strap.

2. The rope with which tackles are *rove* is called a *fall*. To *overhaul* is to separate the blocks. To *round in* is to bring them closer together. When brought together the blocks are said to be *chock*.

3. A tackle is rove by two men, back to back, 6 feet apart; the blocks should be on their sides between the men's feet, hooks to the front and the coil of rope to the right of the block at which there are to be the greater number of returns. Beginning with the lowest sheaf of this block, the end of the fall which is to be the standing end is passed successively through the sheaves from right to left and then made fast.

4. In using tackle great care must be taken to prevent it from twisting. The best method is to place a handspike between the returns, close to the movable block, with a rope to each end, by means of which it can be steadied. New rope must be uncoiled and stretched before using it as a "fall."

Crane chain, when used as a fall, should be thoroughly soaked in oil.

75. TACKLES.

(*See also* G.A.T., Vol. III, Chap. 7.)

1. Various tackles are shown in Pl. 43.

Theoretically, in any system of two blocks the power required to raise a weight W is W ÷ number of returns at the movable block. Owing to friction 10 per cent. must be added to the power for each sheaf used.

If P be the power required to raise a weight W, G the number of returns at the movable block, and N the total number of sheaves; then

$$P = \frac{W}{G} + \left(\frac{N}{10} \times \frac{W}{G} \right)$$

2. The fall, in lifting heavy weights, can rarely be worked by hand, but has to be "led" to either a capstan or winch, by which power is gained and a steady pull ensured.

3. In using tackles with sheers, gyns, or derricks, the running end of the fall should always be led through a "leading block," lashed, as a rule, to one of the spars a few feet above the ground; a snatch block is most convenient for the purpose. (Pl. 44, Figs. 4 and 5.)

76. DERRICKS.
(See also G.A.T., Vol. III, Pt. 3.)

1. A derrick (Pl. 44, Fig. 4) is a single spar set up with four guys at right angles to one another, secured to the tip with clove hitches. (Pl. 42, Fig. 6.) A block for the tackle is lashed to the head, and the derrick can be used for raising and swinging a weight into any position within its reach, which is about one-fifth of its height. The anchorages for the guys should be at a distance from the foot of the derrick equal to twice its height. The foot should be let into a hole in the ground to prevent it slipping and should rest on a bearing plate if the ground is soft.

2. A swinging derrick (Pl. 45, Fig. 4) is an inclined spar with its butt secured near that of a standing derrick and its tip lashed to the main tackle of the latter, and also steadied by at least two side guys. Its head can then be raised or lowered and it may be swung to right or left, *provided that the range of its swing does not exceed 60° altogether.* The detail of the lashing at the butt is shown in Pl. 44, Fig. 5.

3. To raise a derrick, the spar is first laid on the line joining the footing and one of the guy anchorages, with the butt nearly over the footing. A footrope is secured to the butt and to a holdfast on the same side of the footing as the spar is on, and close to it. The four guys having been made fast to the tip and passed to their holdfasts, the tip is lifted as high as possible by hand. The back guy is then hauled on and the fore guy let out until the derrick is in the desired position.

4. In carrying a spar, the party should be equally divided on either side of it, and sized from one end. The spar should then be lifted in two motions on to the inner shoulders of the men. In lowering a spar, the party should slowly face inwards, and lower first the butt end, and afterwards the tip. One man should always give the word for lifting and lowering.

5. The number of men required to erect derricks to lift weights up to 4 tons is from 20 to 25.

77. SHEERS.
(See also G.A.T., Vol. III, Pt. 3.)

1. Sheers require only two guys—"fore" and "back." They should be fastened to the legs above the crutch by clove hitches, the back guy to the fore spar, and *vice versâ*, so that their action may tend to draw the spars closer together and not strain the lashing. The minimum distance of the anchorages from the legs should be double the height of the sheers. The upper block of the tackle is hooked to a sling of rope or chain passed over the crutch. Sheers can, as a rule, be used for heavier weights than derricks, but can only move them in a vertical plane passing between the legs. The feet of sheers must be secured or let into holes in the ground. The distance apart of the legs should not be more than one-third the length of the leg up to the crutch, and the sheers should not be heeled over more than one-fifth of their height.

2. In order to lash the legs they are laid side by side on a skid, and kept 2 inches apart by a wedge. The lashing is commenced with a clove hitch on one spar, carried six or more times upwards round both spars without riding, then two frapping turns, and finished off with two half-hitches round the other spar. (Pl. 44, Figs. 1 and 2.)

CHAPTER XII.—COMMUNICATIONS.

(*See also* Military Engineering, Pt. I, Sec. 15, and Pt. V, Sec. 9.)

78. GENERAL INSTRUCTIONS.

1. Communications of a temporary nature are usually required:—
   (*a*) In connection with a defensive position to enable troops or guns to be readily moved from one portion to another.
   (*b*) For movement across country devoid of suitable tracks.
In both cases provision will generally have to be made for wheeled vehicles.

2. The work in connection with such communications, will usually consist in improving existing tracks, bridging, or filling up soft places, cutting ramps in steep ground, cutting gaps through fences and clearing roads or paths through woods.

3. Within a position, troops should be able to move on as broad a front as possible, and troops and messengers should be guided to their destination by signposts, by "blazing" trees or other means.

4. A roadway 10 feet wide (8 feet minimum) will take a single line of wagons * passing in one direction, or infantry in fours; 12 feet is better for allowing horsemen to pass without difficulty: for each additional line of vehicles 8 feet should be added to the width of the road.

Where there is little traffic, a width of 10 feet may suffice for wagons both going and coming, provided sidings are made at intervals, into which one wagon may go to allow another to pass. Gaps in walls, hedges, etc., forming road boundaries, should be made at least 15 feet wide if intended for wheeled traffic.

A width of 6 feet is sufficient for infantry in file or pack animals moving in one direction.

5. The gradient for a short distance, such as a ramp leading on to a bridge, may be $\frac{1}{3}$ or even $\frac{1}{2}$ for infantry, and $\frac{1}{4}$ for artillery, provided it is straight, but for animals or wheeled traffic slopes steeper than $\frac{1}{10}$ are inconvenient, and if an incline is a long one its slope should be at least $\frac{1}{20}$ for prolonged traffic.

6. When a new road has to be constructed, and time is of great importance, it should be made as straight as is consistent with the extreme gradient permissible. In laying it out, the centre line should be marked by pickets, or the margins by *spitlocking*, and some kind of pathway cleared throughout as an initial measure. The more difficult portions *must* be dealt with first, or the road may remain valueless on the completion of the remainder; and the whole road rendered passable by artillery before any portion is still further improved.

---

* The width over all service vehicles varies from 6 feet (Telegraph Cable Cart, Mk. II) to 7 feet (Ambulance Wagon, Mk. VI), and 2 wheel tracks, each 1 foot wide, spaced at 5 feet apart between inside edges, will accommodate all service vehicles.

If the road passes through a wood, a line of trees in the required direction should first of all be cut down. The space to be occupied by the road should then be cleared and the tree roots grubbed up. Side drains must be cut and relief for storm water passing under the road, if necessary, must be made with the natural line of drainage of the country.

When ascending a hill by means of zig-zags the road should be made as level as possible at each angle, and half as wide again as in the straight portions. The road should be prolonged up-hill about 12 yards beyond each turn to enable teams to pull straight until the vehicle reaches the level turn. Short zig-zags should be avoided, and no curves should be sharper than with a radius of 60 feet for traffic of all arms to meet and pass without producing a deadlock.

7. The best foundation for a temporary road over boggy ground is one or more layers of fascines or hurdles; the top row must lie across the direction of the traffic touching one another. When time is short or suitable material is not at hand, much can be done by throwing down brushwood, heather, or even straw or grass, care being taken that this, like the fascines, is laid across the road (Pl. 50, Fig. 2). It is quite useless to place stones or earth in small quantities upon a yielding foundation.

If much wheeled traffic is expected a reserve of material should be collected at suitable points to replace any that gets worn through.

8. Where timber is available and heavy traffic is expected, a "corduroy" road may be made. This is constructed by felling trees, cutting them roughly to the required lengths and laying them across the road at right angles to its direction, ribands being spiked to them at either end at the required width of the track; or the logs may be held together by interlacing with rope or wire.

The interstices between fascines, brushwood, logs, &c., may be filled with small stones and earth to make a better surface.

9. For the regulation of traffic on military communications, see F.S. Regs., I, Sections 32 and 33.

CHAPTER XIII.—BRIDGES AND THE PASSAGE OF WATER.

(*See also* F. S. Regs., Pt. I, Sec. 32, and Military Engineering, Pt. III.)

(*For materials and stores, see Appendix III, Tables* 5, 6, 9, 12 *and* 14.)

79. GENERAL INSTRUCTIONS.

1. Tactical requirements will usually determine the locality for a military bridge, but when choosing the exact site attention should be paid to the nature of the banks and approaches, the nature of the bed, width to be bridged, depth of water, strength of current, and the probability and extent of floods, all of which are important from a technical point of view. If a tidal river, the rise and fall of the tide should be ascertained.

2. THE APPROACHES AT BOTH ENDS OF A BRIDGE ARE A MATTER OF GREAT IMPORTANCE. EASY ACCESS AND A DIFFICULT EXIT ARE LIABLE TO CAUSE CROWDING AT THE ENTRANCE TO, AND ON, A BRIDGE, WHICH MAY LEAD TO ACCIDENTS AND DELAY.

The passage of troops *off* a bridge should always be expedited, while their passage on to it should be carefully regulated, and, when necessary, checked by material obstacles.

3. Marshy banks should be avoided, and if ramps are required the gradients should be easy. (*See* Sec. 78.)

4. River bends are not, as a rule, good positions for military bridges as the current runs unevenly, the depth varies, and the bank on the out-side of the bend is often precipitous, while that on the inside may be marshy.

Thus, in Pl. 46, Fig. 1, the depth will probably be above the average near C and F, and there will be shallow spits at D and E.

For this reason a river which is not anywhere fordable straight across may be found passable in a slanting direction between two bends, as at A B, Fig. 1, through shoal water.

5. All fords should be clearly marked by strong pickets driven into the river bed above and below the ford, their heads being connected by a strong rope which is securely anchored to holdfasts at each shore end. Marks should be made on those pickets which stand in the deepest water, at a height of 3 feet and 4 feet above the bottom, in order that any rise of water above the fordable depth may be at once evident.

6. The simplest plan for measuring the velocity of a stream is described in Sec. 61 (2). Seven-tenths the mean number of feet a second gives the number of miles an hour, in which terms the velocity should be stated.

80. Bridging Expedients.

1. For tactical reasons it is usually important to pass men, and often horses and artillery, across water at the earliest possible moment, and, in the absence of a bridging train, some expedient, more rapidly executed than the construction of a bridge for all arms, must precede work of a more deliberate nature.

2. It will frequently happen also that materials for the construction of a bridge will be lacking in whole or part. In such a case the means of crossing a river must be improvised from the materials likely to be found near at hand or which are usually carried by the troops. A few of these bridging expedients are given below:—

(i) In shallow water, carts or wagons may be used to form the substructure for a bridge.

(ii) Small gaps may be filled up with bundles of brushwood, channels being left for the passage of the water.

iii) Rafts or even piers for bridges may be made of waterproof material such as tarpaulins, ground-sheets, &c., stuffed with hay, straw, heather, ferns, &c.

A raft consisting of four 18 foot by 15 foot tarpaulins (*see* App. III, Table 11) stuffed with hay will carry a G.S. wagon, 18pr. field gun without limber, or equivalent load not exceeding 24 cwt. The best method of filling each tarpaulin is to make a light framework of poles, 6 feet square by 2 feet 6 inches high, on the ground (a hole of similar dimensions will do almost as well). Then place two lashings about 24 feet long across the framework each way, and over these the tarpaulin, well-soaked. Fill the tarpaulin with hay and trample it well down. The ends and sides of the tarpaulin are then folded over the hay, and the whole made into a compact bundle by securing the lashings across the top. (Pl. 47, Fig. 1.)

Two of these floats are then lashed together by means of two 14ft. spars. This forms half the raft. The other half is made in a similar manner. The two halves are then lashed to one another, and 8ft. apart, by means of four 16ft. roadbearers. The raft will then measure 15ft. by 12ft. by 8ft. 6in. In such a raft the buoyancy is greatly in excess of that actually required to carry the load, but this is necessary owing to the kind of material employed and the short length of the piers. With good tarpaulins the buoyancy will remain good for at least 8 hours.

The stores required are:—

Tarpaulins	4
Hay (tons)	1½
Planks	16
Spars (average 4″ diam.), four 16′, four 14′, and two 12′	10
Lashings, 1″, about 8 fms. long	40
Do. 1½″, about 6 fms. long	16
Ropes, 2″, length according to width of river	2
Punting poles	2

Smaller rafts can similarly be made by stuffing ground sheets with hay or straw. 24 of these made into a raft will support a load of 1,800 lbs. (Pl. 47, Figs. 3 and 4.)

(iv) A rough boat can be made by covering the body of the Mark IX G.S. wagon with its tarpaulin cover. Any projecting points of the wagon must be covered with hay to protect the tarpaulin, and any holes in the wagon should be filled in the same way. The tarpaulin must be kept close to the wagon body by lashings. If sufficient lashings are available, one should be tied right round the wagon two-thirds of the way back from the front, while a second should be lashed round the wagon half way up its side, passing through the eyelet holes in the cover. 4 to 6 men may be carried sitting down in this boat.

Rough boats may be made in a similar way by fastening tarpaulins over a brushwood framework.

(v) When casks are obtainable in small numbers only, light foot-bridges may be made in various ways.

A bridge of this description with a double footway, suitable for infantry, can be quickly constructed as shown on Pls. 49 and 50. The stores required for each raft or 18ft. of bridge are :—

Casks, 54 gallons 4
Gunnel spars, 15 ft. long, about 8 ins. diameter ... 4
Tie-baulks ditto ... 4
Planks, 10 ft. long 8
Lashings, 2 ins., 9 fms. 8
ditto 1½ ins., 6 fms. 4

This bridge will bear infantry in single file at intervals of 4 feet on both footways at the same time. As the footways are at the ends of the piers, the bridge is suitable for swimming animals alongside. (Pl. 50, Fig 1.)

(vi) A raft which will carry two armed men can be made from a single 54 gallon cask, as shown in Plate 48, Fig. 3.

81. TACKLE FOR SWIMMING HORSES.

A method of swimming horses across a river where no bridge exists is by means of an endless rope.

If possible, a place should be selected where the banks shelve and the bottom is hard on each side of the river: this is almost essential for the landing side.

The materials required are a two-inch rope (endless), the minimum length of which should be equal to twice the width of the river plus about 70 yards; pickets for holdfasts, four snatch blocks and a tackle.

A party of 1 officer and 35 men are sent across the river with 2 snatch blocks, the tackle, pickets for 2 holdfasts and a bight of the endless rope. This is arranged as shown on Pl. 46, Fig. 2. 10 to 15 men are required on the starting side to work the rope and four men as lashers. Each horse having been stripped, except for head collar and head rope, is fastened by one of the lashers, walking backwards, to the endless rope on the down stream side by means of a draw hitch

(Pl. 41, Fig. 3), made with the head rope, the length of which must not exceed 18 inches when the horse has been secured. The horses should be tied on at intervals of 3 or 4 horse lengths (8-10 yards), and care must be taken that the head rope is tight on the endless rope. On the landing side 25 men will be required to pull on the rope which they should do hand over hand and as quickly as possible. Four men with clasp knives receive the horses, and must be ready to cut any head rope knot that jams. Six men take the horses from the receivers and lead them away.

If snatch blocks are not available more men will be needed, the horses should only be sent across two or three at a time, and a spliced rope is not essential.

(For making a *long splice*, see G.A.T., Vol. III.)

82. TYPES OF BRIDGES.

1. Where the tactical situation demands better facilities for the passage of water or ravines than those existing or above described, and time, materials, and labour are available, some form of bridge may be required.

The type of bridge will vary according to the materials available, the traffic expected and the nature, breadth, depth, &c., of the span to be bridged: but WHATEVER ARM OF THE SERVICE A BRIDGE IS CON-STRUCTED TO CARRY, IT SHOULD BE CAPABLE OF CARRYING THAT ARM WHEN CROWDED.

A bridge that will carry infantry in fours crowded will carry field guns and 5-inch howitzers and most of the ordinary wagons that accompany an army in the field.

2. The officer superintending the construction of a bridge is responsible that it is strong enough to support the weight it is intended to carry. To prevent it being improperly used, and in addition to the measures prescribed in F.S. Regns., Pt. I, Section 32, he should place a sign board at each end, stating the greatest permissible load, thus:—

"Bridge to carry infantry in fours."
"Bridge to carry infantry in file."
"Bridge to carry guns not heavier than 18-pr."
"Not for animals."
"Bridge for All Arms. No Road Engines."

The simplest form of construction consists of round spars lashed together with rope or wire, but sawn timbers, as used in the construction of buildings, and iron fastenings, are often more easily obtained than spars and rope. Iron fastenings necessitate a few carpenter's tools, but are more satisfactory than lashings.

The *length* of bays depends chiefly upon the strength and length of the materials available as roadbearers. When the piers are high, or involve much time in construction, material and labour should be economized by making the bays as long as possible.

The *number* of roadbearers also depends upon the size of the timber available. (*See* Section 95.)

3. When the bottom can be reached throughout, a trestle bridge (Pl. 51, Fig. 5), or some form akin to it, will generally be the most economical in material and the easiest to make. When the bottom

cannot be reached small spans may be bridged by laying timber straight across the gap, or by using such forms of bridge as the single lock, the double lock and the cantilever. Where material is available and depth of water and current are suitable, a floating bridge will be the quickest and simplest form of bridge to employ.

83. ROADWAYS.

1. The same nature of roadway can be applied to each type of bridge, and its usual form is shown in Pl. 51, Fig. 1.

The planks or *chesses*, A, A, placed across the width of roadway are supported on longitudinal *baulks* or *roadbearers*, B, B, which in their turn rest on transverse *transoms*, T, T. The method of supporting these transoms depends on the type of bridge. The chesses are kept steady by *ribands*, R, R, which are secured to the outside baulks either by *rack lashings* or by lacing, or the chesses may be nailed down.

2. The roadway is generally constructed with a slight rise towards the centre of the bridge to get loads on to the bridge quietly and off it easily; this is technically called the *camber*, and should be about $\frac{1}{30}$ of the span for bridges up to 40 feet in length. For longer bridges, camber at a slope of $\frac{1}{30}$ for the bays next the banks is sufficient. In soft ground an extra allowance for subsequent settlement should be made.

3. The "normal" width of bridge is 9 feet *in the clear—i.e.*, the clear space between the ribands. As a rule this should not be exceeded. A width of 8 feet in the clear will suffice for infantry in fours, military vehicles and cavalry in half sections.

Six feet with no handrail posts will take infantry in file, cavalry in single file, and field guns passed over by hand; $1\frac{1}{2}$ to 3 feet will take infantry in single file.

4. Chesses or planks $1\frac{1}{2}$ to 2 inches thick are sufficient for ordinary traffic provided that they are supported at intervals not exceeding $2\frac{1}{2}$ feet; if they are longer than the width of the bridge they can be economized by placing them diagonally or longitudinally. Planks 10ft. long by 2 inches thick, and supported at the ends only, must not be marched over by more than two armed men at the same time.

For continuous or heavy wheeled traffic additional chesses should be laid longitudinally, to form wheel tracks.

Hurdles, short fascines, corrugated iron, &c., can be used in lieu of planks, but are not good for horse traffic.

When material is available, chesses should be laid on the ground on the banks on each side for a short distance, to allow horses to become accustomed to the noise before actually getting on to the bridge, and to prevent the ground becoming cut up.

5. In most bridges the ribands should be fairly pliant, in order that the rack lashings may press them tightly down on the chesses throughout. In floating bridges, however, stiff ribands are desirable, as they tend to stiffen the bridge; for the same reason trestles should be braced diagonally to those on either side (Pl. 51, Fig. 5).

Rack lashings should be applied at intervals of 4 feet or 5 feet.

6. Handrails should be provided, especially for horse traffic. The normal height above the road is 3 feet. They must be strongly built. The handrail posts must not be nearer than nine inches from the inner

edge of the ribands for wheeled traffic. Screens on each side are desirable for passing animals over a bridge, especially over running water, while straw or rushes in the absence of moss litter, tan, or sawdust may have to be spread on the chesses to prevent slipping. If earth is used it must not be less than 3 inches thick, and allowance must be made in the calculations for the extra weight. (Sec. 98.)

7. If the slope of a boarded roadway, intended for use by animals, exceeds $\frac{1}{10}$, light battens must be nailed across the roadway at intervals of about 1 foot to afford foothold in wet weather, the wheel tracks being left clear.

8. All bridges require firmly secured shore ends. A section of a normal type is shown in Pl. 51, Fig. 4.

84. TRESTLE BRIDGES.
(See also Military Engineering, Pt. III, Sec. 7.)

1. Before proceeding with the construction of the trestles a complete section of the gap to be bridged showing the water level, if any, should be marked out on a piece of flat ground. To lay out this section, the width of the gap and its depth to firm bearing at the proposed position of each trestle must be measured along the centre line of the bridge. A long light pole with a string and weight are useful for this purpose, or the method shown in Pl. 32, Fig. 4. If the bottom is very uneven a section for each side of the bridge will be necessary. For a muddy bottom the ledgers should be close to the butts, so as to take the mud ; for a rocky bottom they should, if used, be high enough up not to touch the bottom.

2. Trestles may be made in various ways, some of which are illustrated on Pl. 52. The best are those put together with iron fastenings, then come those with wire lashings, while those with rope lashings only are the least satisfactory owing to the impossibility of keeping the lashings taut.

3. Iron fastenings may consist of dogs, spikes, drift-bolts, nails, bolts, chains, &c. With dogs the position of each must be chosen with the definite object of preventing a possible distortion of the frame. They should be on both sides of the frame. Dogs should not be driven within 3 inches of the edge or 4 inches of the end of a timber.

Spikes, when driven in pairs, should incline towards each other. They generally run from 5 to 10 inches in length. (See Appendix III, Table 9.) Spikes with chisel points should be driven so that the edge is across the grain.

Drift-bolts are made of round iron, pointed at one end and with a small head at the other. They may be of any length, and are especially useful for fastening horizontal timbers to the top and bottom of upright ones. Holes slightly smaller than the bolts should be bored to receive them. Wooden trenails may sometimes be used in place of drift-bolts.

4. In making trestles with wire or rope lashings, square lashings (Pl. 42, Fig. 4) should be used except where the diagonal braces cross. The transom and ledger should be on the same side of the legs parallel to each other ; while the braces are put on the frame with both tips and

one butt on the opposite side of the legs to the transom, the other butt being on the same side.

5. The butts of braces can be lashed simultaneously with the ledger and transom. The frame must then be *squared* by adjusting it so that the diagonals, measured from the centre of the ledger lashing to the centre of the transom lashing on the opposite leg, are of equal length. The braces can then be lashed at the tips and crossing point.

6. If the timber is weak, both legs and transom can be doubled. It is a great advantage if a direct support can be given immediately below all transoms. The usual methods of supporting transoms are shown in Plate 48, Figs. 4 to 8, and Plates 51 to 54. Ledgers and diagonal braces may be of light material, as little strain is brought upon them, but they should be well secured.

7. When the water is very shallow trestles can be carried out and placed by men working in the water. When the water is too deep for this they can be carried on to the bridge and lowered feet first down inclined spars to their final position (Pl. 51, Fig. 5), or taken out on rafts and tipped up into position by means of guys.

8. Trestles, of which the legs are all in one plane, are kept vertical by fastening the roadbearers to the transoms and by cross-bracing from each trestle to its neighbour (Pl. 51, Fig. 5). The nearest trestles to the banks on either side should also be rigidly connected to bollards on the bank by light spars fastened to the tips of the legs. These light spars are put on before the trestle is launched, and help to get it into position, and must be secured before the first bay is used for placing the second trestle.

9. On trestle bridges with iron fastenings handrails may be fixed as shown on Pl. 52, Fig. 1. On trestle bridges where lashings are used they may be fastened to the tips of the legs.

10. Pl. 51, Fig. 2, shows a four-legged trestle made of two frames similar to two-legged trestles, locked at the transoms, and connected by short ledgers at the feet. One frame must therefore be narrower than the other. The inclination of the legs should be such that the breadth of the base on which the trestle stands should not be less than half the height. The legs must also have an outwards splay of $\frac{2}{9}$.

Four-legged trestles can be made of fairly light material, and will stand without bracing. They are consequently useful for small bridges of two bays, requiring one central support, and as occasional steadying points in a long bridge of two-legged trestles: but they are more difficult to place than the latter.

Pl. 52, Fig. 5, shows a four-legged trestle made of squared timbers spiked together.

A light and rapidly constructed trestle footbridge suitable for infantry in single file is shown in Pl. 48, Figs. 1 and 2.

11. One of the most useful substitutes for trestles is cribwork (Pl. 51, Fig. 3), when timber is plentiful and other stores deficient. If used in water a tray should be formed in the bottom of the crib, which can then be towed into position, weighted with stones and sunk.

When timber is scarce and stones plentiful, " crate " piers may be made of stones enclosed in an open framework of timber.

12. The presence of a pier in running water previously unobstructed causes an underscouring action by the water to commence on the upstream side of the pier, which may eventually capsize the pier.

This can be temporarily guarded against by surrounding the upstream side of the pier with boulders or sacks of small stones.

85. Cantilever Bridges.

1. Pl. 53, Fig. 1, shows a type of cantilever bridge.

For all spans the method of construction is practically identical. A site is chosen where a large rock or rocks rise out of the stream, or a pier is constructed of dry stone work and wooden bindings. On the top of these are laid a number of stout beams (*aa*), projecting over the stream, with the projecting ends somewhat higher than the shore-ends. The number of beams, their length and amount of projection depend on the span.

Should the remaining span be too large for available timbers, another row of cantilevers can be placed on the first row *aa*, and two more transoms *tt* near the projecting ends and so on. There are more cantilevers in the bottom row than in the row above and so on.

The step from the top row of cantilevers to the top of the central roadbearers can be avoided by lashing the top transom underneath the ends of the top row of cantilevers instead of on the top, or an extra row of roadbearers may be added above the top row of cantilevers.

2. The shore ends of the cantilevers must be firmly anchored. This may be done (1) by piling earth and stones on the shore ends as a counterweight: boards, if available, being nailed across the tops of each layer to give a larger surface for the material to bear upon ; (2) by excavating an anchor trench vertically below the shore ends of the cantilevers, and anchoring them down with wire to a log placed at the bottom of the trench, which is afterwards filled in.

Each row of cantilevers should also be lashed to the row below.

86. Frame Bridges.
(*See also* Military Engineering, Pt. III.)

1. Frame bridges are of two types, the single lock and the double lock, but it will only be on rare occasions that they will be required, such as replacing a broken span in a viaduct with high piers.

The single lock bridge (Pl. 53, Fig. 2) consists of two frames, one of which is narrower than the other so that they may lock together. It supports one central transom and is sufficient for spans up to 30 feet. 3 N.C.Os. and 30 men should be able to make one of these bridges in about 1 hour.

The double lock bridge (Pl. 53, Fig. 3) consists of two frames of the same width which are held apart by distance pieces, and support two transoms. It does for spans up to 45 feet. 2 or 3 N.C.Os. and from 30 to 40 men should make one of these bridges in about 3 hours.

2. The span of a frame bridge is the horizontal distance between the footings of the frames, and is independent of any increase of span due to sloping banks or bays of trestles.

3. The frames are nearly identical with two-legged trestles (Pl. 52, Fig. 4), but the slope of the legs is not so great $\frac{3}{20}$ generally sufficing, and the transom and ledger are lashed to opposite sides of the legs, transom on the shore side so as to bear on the legs, and ledger on the opposite side so as not to interfere with the footings. Before lowering the frames into their places, footings must be prepared, holdfasts driven, fore and back guys attached to the top of each leg, and foot ropes to the bottom of each leg below the ledger. The frames are then lowered and if a single-lock bridge, locked; if a double-lock bridge, held back by the guys a little higher than their ultimate position. A single-lock bridge is then completed with the usual roadway; for a double-lock bridge two distance pieces must be placed across the ends of the frame transoms, and the roadbearing transoms lashed across, as shown in the diagram. The back guys can then be eased and the bridge allowed to lock. The roadway is completed as usual.

4. In order that the parts of frame bridges may fit together, considerable accuracy is necessary in taking the measurements and marking the positions for the lashings. To this end a section of the gap and proposed bridge should be marked out on the ground, allowing for camber. The spars for the legs should then be laid on this section in the exact positions they will occupy when in bridge, and marked to show the proper spots for lashing on the ledgers and transoms.

5. The following are the approximate dimensions of timbers for single and double lock bridges to carry infantry in fours crowded :—

Legs ... ... 7 inches at tip.
Frame transoms, mean diameter ... ... 6 inches.
Distance pieces „   „   ... ... 11 inches.
Other spars as for trestle bridges.

6. When a few nails are available the single lock bridge can be greatly simplified by utilizing two bridging planks for the central transom as shown in Pl. 53, Fig. 2(A).

The frames in this case are of equal width, and may consist of several light legs in place of two heavier ones.

87. FLOATING BRIDGES.

(See also Military Engineering, Pt. III.)

1. In selecting a site for a floating bridge, a spot should be chosen where good holding ground for anchors is available.

Material may be economized by making use of islands.

The roadway of floating bridges is similar to that already described in Sec. 83.

2. Each pier must have enough available buoyancy to support the heaviest load that can be brought on to one bay of the bridge. No extra allowance of buoyancy need be made if the load is live.

The length of each pier should be at least twice the breadth of the roadway for the sake of steadiness, and with the same object they may be connected together at their ends by tie baulks (Pl. 49, Fig. 1) and diagonally stiffened with lashings (Pl. 56, Fig. 1).

The waterway between the piers should never be less, and should if possible be more, than the width of those piers.

Piers may be made from specially constructed pontoons, boats, casks, timber rafts, or inflated skins, or the methods described in Sec. 80.

3. A bridge can be put into position in the following ways:—

(i.) By *booming* out, *i.e.*, when the head of the bridge already constructed is continually pushed out into the stream, fresh materials being added at the tail. This method cannot be used with steep banks and deep water close in-shore.

(ii.) By *forming up*, *i.e.*, when material is continually added to the head of the bridge, the tail being stationary. This method is uninfluenced by the nature of the banks, no men being required to work in the water. Its only drawback is the distance the roadway materials have to be carried.

(iii.) By *rafting*, *i.e.*, when the bridge is put together in different portions, or *rafts*, along the shore, each raft consisting of two or more piers, and these rafts are successively warped, rowed, or towed into their proper positions in the bridge.

This method has the advantage that a large number of men can be employed simultaneously; and, if secrecy be an object, the various portions can be constructed at some distance from the eventual site of the bridge, and a favourable opportunity seized for its construction.

(iv.) By *swinging*, *i.e.*, when the entire bridge is constructed alongshore, and then swung across with the stream.

A long bridge can be constructed by a combination of two or more of the above methods.

4. If a bridge has to remain down for some time, arrangements may have to be made for the passage of river traffic. This can be done by having two or more rafts, at the centre of the bridge, arranged for "forming out"; or the two halves of the bridge may be swung, to afford the requisite passage.

5. Arrangements must always be made, up-stream, for the protection of a bridge from damage by floating substances, either by a boat patrol or by posting men at each pier to pole off such floating objects into the fair way.

6. If heavy siege artillery has to be passed over a broad river it will generally be most economical of material to construct the bridge of only sufficient strength for the ordinary traffic, and to warp the guns across on specially constructed rafts.

88. PIERS OF BOATS.

(*See also* Military Engineering, Pt. III.)

1. The available buoyancy of a boat may be determined by loading it with unarmed men to within about 12 inches of the gunwale, and multiplying this number by 160. The result gives the available buoyancy in pounds. A still larger limit of safety will be required in rough water or a violent current.

2. Boats, which are to be used as piers of a bridge, should be placed "bow on" to the current, and slightly down at the stern; if the river is tidal, they must be placed pointing up and down stream alternately.

8. If the boats are not each buoyant enough to form a pier, they may be used in pairs (Pl. 54, Fig. 8). The sterns are lashed together, and the spars AA_1 BB_1 are held over the side; four 2-inch ropes at AB, CD, C_1D_1, A_1B_1, are passed under the boats and secured to the poles, and four double ropes are passed round the latter at the same points and across the boats; these ropes are racked up tight. Crosspieces, MM, are then lashed to the poles and thwarts, and blocks on the thwarts at EE support the saddle beam, which is lashed to the thwarts and to the stern rings of the boats.

4. Few boats, with the exception of heavy barges, are strong enough to allow of the baulks resting direct on their gunwales. A central transom should be improvised, which can generally be done by resting a beam on the thwarts, and blocking it up from underneath, thus bringing the weight directly on to the kelson. This arrangement is shown in Pl. 54, Fig. 4

89. PIERS OF CASKS.

(See also Military Engineering, Pt. III.)

1. The following detail shows the most convenient way of making a cask pier with large casks (Pl. 55).

The number of men required is two more than double the number of casks, or $2n + 2$ where n is the number of casks. (See Sec. 97.)

The casks are first laid bung uppermost and aligned: and then two baulks, technically known as gunnels, (GG), are placed over the ends of the casks by 4 men, while the remainder of the men stand opposite the intervals between the casks on either side.

The gunnel men at one end place the eyes of the slings (SS) over the gunnels; the gunnel men at the other end secure the slings to their ends of the gunnels with a round turn and two half hitches (Pl. 40, Fig. 5). The brace men keep the slings under the casks with their feet, and, as soon as they are secured, adjust the braces as follows, working simultaneously by word of command :—

The eye of the brace is passed under the sling in the centre of the interval between two casks, the end passed through the eye and hauled taut, the sling being kept steady with the left foot. The brace is then brought up outside the gunnel immediately over the eye, and a turn round the gunnel taken *to the left*, the foot is removed from the sling, and each man then hauls up the standing part of his brace with the left hand, holding on to the turn with the right; as soon as the brace is taut the turn is held fast with the heel of the left hand, and the remainder of the brace, in a coil, is placed on the cask *to the left*. Each man then takes his opposite neighbour's brace from the cask on the right, and passes it between the standing part of his brace and the cask *on his left*, then back between his brace and the cask on his right, keeping the bight so formed *below* the figure of 8 knot on his own brace, and placing the end on the cask to his right. Each man then takes back his own brace from the cask on his left, passes it under the gunnel *to the left* of the standing part, places his foot against the gunnel, and hauls taut. The pier is then rocked backwards and forwards, all the brace men taking in the slack of their braces and hauling taut until the word *steady* is given, when they take a round turn round the gunnel *to the left* of the previous turns, and make fast with two half-hitches round the two parts of their own brace close to the gunnel, drawing the two parts close together and placing the spare ends of their brace between the casks. The pier is then turned up on one side, and the sling adjusted below the third hoop of the casks, and a breast line attached to a sling at each end : it is then lowered and turned up on the other side, the other sling adjusted, a sledge technically called the *ways*, brought up into position, and the pier lowered on to it ready for launching.

2. For a pier of the size shown in the figure (the casks used being butts) the following are needed: *Gunnels*, 21 feet by 4 inches by 5 inches ; the *slings* of 2½-inch rope, 6 fathoms long, with an eye splice 1 foot long at one end ; *braces* of 1½ inch rope, 8 fathoms long, a small eye splice at one end, and a figure of 8 knot 1 foot 5 inches from the eye.

3. Pl. 55, Fig. 4, shows how piers can be made with small casks. Small piers of three or more casks, *aa*, *bb*, *cc*, are first made as described above, and subsequently united by two large gunnels, X, X.

4. Headless casks must be enclosed vertically in a specially prepared framework.

The ends of cask piers should always be rigidly connected to each other by *tie baulks*, which must be lashed to *both* gunnels of each pier ; the roadway baulks can then be laid, without lashing if rectangular, so as to rest on *both* gunnels of each pier. If there is likely to be much sway on the bridge, *e.g.*, on a bridge for animal traffic, or if round baulks are used, some of them should be lashed both to each other and also to the gunnels.

90. RAFTS AND PIERS OF LOGS.

(*See also* Military Engineering, Pt. III.)

1. To form a log raft, the logs should be placed side by side thick and thin ends alternating; they should then be strongly secured with rope, and if possible, by cross and diagonal pieces of timber fastened by spikes or wooden trenails; or the logs can themselves be connected by dogs.

2. If a log raft is to be used as a pier in a bridge, it will frequently be necessary to place the logs in two layers, to avoid obstructing the waterway. A central raised transom must be used. The up-stream end of the raft may, with advantage, be slightly convex.

Rafts are most easily put together and manipulated in the water.

3. It is essential for the sake of stability, that, unless the buoyancy is much in excess of that actually required, the length of each pier of a raft should be twice the width of the platform of the raft. IF THE RAFT IS FORMED OF ONE PIER ONLY THE PART OF THE PLATFORM LOADED SHOULD ONLY BE THE CENTRAL QUARTER (Pl. 55, Fig. 6).

91. ANCHORS AND CABLES.

(*See also* Military Engineering, Pt. III.)

1. For ordinary bridge work 56-lb. anchors, with a reserve of 112-lb. anchors, will generally suffice for moderate streams. The following substitutes may be employed :—Two or more pickaxes lashed together ; heavy weights, such as large stones or railway irons (the latter are best when bent) ; or sacks, &c. filled with stones.

2. The cables are generally of 3-inch cordage. The length of cable "out" should be ten times the depth of the stream, and rarely less than 30 yards. The cable is attached to the ring of the anchor (Pl. 56, Fig. 3) by a fisherman's bend ; a buoy should be attached to the anchor by a buoyline of 1-inch rope to mark its position and to serve as a means of tripping it. One end of the buoy line is fastened to a ring of the buoy by a fisherman's bend, and the other round the crown of the anchor with a clove hitch split by the shank, and two half-hitches round the shank.

3. As a rule there should be an up-stream and down-stream anchor to every second pier of a floating bridge.

If anchors are scarce, one may be made to serve for two piers by attaching two cables to it on the down-stream side of the bridge, or when only acting as wind anchors in still water.

Care must be taken before *heaving* an anchor overboard to see that the *stock* has been properly keyed to the shank.

Timber rafts and cask piers bring a greater strain on anchors than boats or pontoons.

4. In a very rapid current, anchors are liable to drag or to pull the piers down by the head. The bridge must then be secured to a hawser stretched across the river up-stream. Wire rope is convenient for the purpose. The piers of bridges can be anchored direct to the banks, up and down stream until the cables make an angle of less than 45° with the roadbearers, or become inconveniently long.

92. FERRIES AND FLYING BRIDGES.

(*See also* Military Engineering, Pt. III.)

1. In the simplest form of permanent ferry, boats or rafts are hauled backwards and forwards from bank to bank by means of ropes stretched across the river. Such a rope should, if attached to the boat, &c., be made fast at the stem or stern and not amidships. If it is not convenient to stretch a rope across the stream on account of traffic or other reasons, or if the current is rapid and regular, a *flying* bridge may be used. This is a form of bridge in which the action of the current is made to move a boat or raft across the stream by acting obliquely against its side, which should be kept at an angle of about 55° with the current. (Pl. 56, Fig. 2.) Long narrow deep boats with vertical sides are the best for the purpose, and straight reaches the most suitable places, as they are generally free from irregularities of current. It is necessary to have a vertical surface for the current to act against. If, therefore, the boat is a shallow one or if the raft is made of casks or

other material with a curved surface, vertical boards called lee boards must be lashed to its side.

2. The cable, which should, if possible, float, can either be anchored in mid-stream or on shore at a bend of the river, and the raft can swing between two landing piers. The length of a swinging cable should be from one and a half times to twice the breadth of the river, and it will work better if supported on intermediate buoys or floats to prevent it from dragging in the water. Telegraph wire, buoyed as above, makes a good swinging cable.

3. Another way is to stretch a wire cable across the river, and arrange for the raft to travel along it by means of a block with large sheave. (Pl. 56, Fig. 4.)

93. Weights of Troops, Guns, and Materials.

1. The maximum weights brought on a bridge by the passage of troops in marching order are:—

Infantry, in file, crowded at a check,	$2\frac{1}{4}$ cwts.	}	per lineal foot of the bridge irrespective
„ in fours „ „	5 „		of width of road-
Cavalry, in single file „ „	$1\frac{3}{4}$ „		way, provided the
„ in half-sections „ „	$3\frac{1}{2}$ „		formation is preserved.

Armed men in a disorganized mass may weigh $1\frac{1}{4}$ cwt. per square foot of standing room.

2. The maximum weights brought on a bridge by guns of an army in the field are:—

18-pr. Q.F. gun, maximum concentrated weight on one bay,	18 cwt.
18-pr. Q.F. gun	do.	do.	24 cwt.
4·5-in. Howitzer	do.	do.	26 cwt.
60-pr. B.L. gun	do.	do.	67 cwt.

The reason for not including the weights of the limbers is that, so long as the distance apart of the axles exceeds a certain proportion of the span, the greatest effect produced on the roadbearers occurs under the heavier of the weights on the two axles when it moves over the centre of the span. (*See* Mil. Eng., Pt. III, Sec. 10.)

3. The weight of earth 1 inch thick may be taken as 0 lbs. per square foot covered; and of soft timber 40 lbs. per cubic foot.

94. Formulæ for Calculating Size for Roadbearers, Transoms and Cantilevers.

1. Good rough formulæ for calculating the sizes necessary for roadbearers and transoms, etc. are given below. The formulæ include a factor of $1\frac{1}{2}$ for live load, in addition to a factor of safety of 3; they also allow for a normal weight of superstructure.

For unselected rectangular beams supported at both ends:—

$$W = \frac{bd^3}{L} \times K \quad . \quad . \quad . \quad . \quad . \quad . \quad (A)$$

For unselected round spars supported at both ends:—

$$W = \tfrac{6}{10} \times \frac{d^3}{L} \times K \quad . \quad . \quad . \quad . \quad . \quad (B)$$

2. The formulæ for strength of cantilevers are as follows:—
For unselected rectangular beams fixed at one end :—

$$W = \tfrac{1}{4} \times \frac{bd^3}{L} \times K \quad . \quad . \quad . \quad . \quad (C)$$

For unselected round spars fixed at one end:—

$$W = \tfrac{1}{4} \times \tfrac{6}{10} \times \frac{d^3}{L} \times K \quad . \quad . \quad . \quad (D)$$

3. In the above formulæ:—
W = actual load on one beam in cwts. evenly distributed (without superstructure).
b = breadth of beam in inches.
d = depth of beam in inches.
$L = \begin{cases} \text{length of beam in feet between points of support for (A) and (B).} \\ \text{length of cantilever in feet from point of support for (C)} \\ \text{and (D); or, if loaded at one point only (\textit{see} below), length} \\ \text{from point of support to position of load.} \end{cases}$
K = a variable quantity for different timbers.

For larch and cedar ... ... $K = 1$
  ,,   Baltic fir ... ... $K = \tfrac{5}{6}$
  ,,   American yellow pine ... ... $K = \tfrac{4}{5}$
  ,,   beech and English oak ... ... $K = \tfrac{3}{4}$

In the case of round spars b and d are identical, and their strength is only about $\tfrac{6}{10}$ that of square beams of the same depth.

4. To use these formulæ for a *concentrated* weight, such as a gun, the total weight on the gun wheels must be multiplied by two to convert it to the equivalent *distributed* weight, when it can be substituted for W. When, as in the case of a transom, the load is applied at several points, it can be taken as distributed.

5. With several baulks under a roadway, the two outer are assumed to carry only half as much of any weight brought on the bridge as the inner ones. Thus, with five baulks, the outer baulks each bear $\tfrac{1}{8}$ total weight, the inner baulks each bear $\tfrac{1}{4}$ total weight. In calculations, the greater weight must be worked to in order that the baulks may be interchangeable if necessary. If roadbearers differing in strength are used, the strongest should be placed beneath the wheel tracks.

6. Rectangular beams should always be used on edge, in order to obtain the maximum of rigidity and strength. In calculating the sizes of beams by the formulæ in 1 and 2 above if b and d are unknown, d should be considered equal to $2b$, but may be selected within the limits $d =$ from $1\tfrac{1}{2}b$ to $3b$ in order to make use of such materials as may be available.

The strength of a composite beam formed by fastening together two or more planks on edge, side by side, may be considered equal to that of a solid beam of the final dimensions, within the limits of b and d given above; but if the planks are used placed one above the other in *horizontal* layers, the total strength of the beam must only be taken as equal to the sum of the strengths of each layer considered separately, and will not increase with the square of the total depth.

A tapering spar, when supported at both ends and overloaded, will break in the centre, and not at the small end. In calculating the strength of such a spar when used as a baulk, d is therefore taken at the centre of the spar.

95. TABLE GIVING SIZES OF ROUND ROADBEARERS AND TRANSOMS FOR VARIOUS LOADS AND SPANS $(K=\frac{5}{4})$.

Loads on Bridge.	Round Spars required	Mean Diameter in Inches.					
Infantry in file.	3 Roadbearers ... each Transoms .. ,,	6 6½	6½ 7	7 7	7½ 7½	8½ 8	9 8
Cavalry in single file.	3 Roadbearers ... each Transoms ... ,,	5 6	5½ 6	6½ 6½	7 7	7½ 7	8 7½
Infantry in fours.	5 Roadbearers ... each Transoms ... ,,	6 9	6½ 9½	7 10	7½ 10½	8½ 11	9 11½
Cavalry in half-sections.	5 Roadbearers ... each Transoms ... ,,	5 8	5½ 8½	6½ 9	7 9½	7½ 10	8 10
Q.F. 13-pr. gun, Marks I and II	5 Roadbearers ... each Transoms ... ,,	5½ 7	6 7	6 7½	6½ 7½	7 7½	7½ 7½
Q.P. 18-pr. gun, Marks I and II.	5 Roadbearers .. each Transoms .. ,,	6 7½	6 7½	7 8	7 8	7½ 8	7½ 8
5″ Howitzer, Mark I	5 Roadbearers ... each Transoms ... ,,	6 8	6½ 8	7 8½	8 8½	8½ 8½	9 8½
60-pr. gun in travelling position.	5 Roadbearers ... each Transoms ... ,,	8 10	8½ 10	9 10	9 10	9½ 10½	10 10½
4.7″ gun on travelling carriage.	5 Roadbearers .. each Transoms ... ,,	8½ 11	9 11	9½ 11	10 11	10½ 11	10 11
	Spans in feet ...	10	12	14	16	18	20

Other timbers not affected by length of bay:—
　Ledgers and handrails, mean diameter, 4 inches to 6 inches.
　Braces and ribands, 3 inches at tip.
The depth (d) and breadth (b) of rectangular timber of equivalent strength to the round spars given in the above table can be found from

the formula $bd^3 = \frac{A}{10} D^3$; where D is the mean diameter in inches given in the table. Some practical equivalents for ready reference are given below :—

D	b by d	D	b by d	D	b by d
3	2 by 3	6	3 by $6\frac{3}{4}$	9	5 by $9\frac{3}{4}$
$3\frac{1}{2}$	2 by $3\frac{3}{4}$	$6\frac{1}{2}$	3 by $7\frac{1}{2}$	$9\frac{1}{2}$	5 by $10\frac{1}{2}$
4	2 by $4\frac{3}{4}$	7	3 by $8\frac{1}{4}$	10	5 by $11\frac{1}{4}$
$4\frac{1}{2}$	2 by $5\frac{1}{4}$	$7\frac{1}{2}$	3 by 9	$10\frac{3}{4}$	5 by 12
5	3 by $5\frac{1}{4}$	8	4 by 9	11	6 by 12
$5\frac{1}{2}$	3 by 6	$8\frac{1}{2}$	5 by 9	12	7 by 12

96. SIZE OF TIMBER FOR TRESTLE LEGS, ETC.

From the tables on Plates 57 and 58 the sizes of timbers required for trestle legs, derricks, &c., when exposed to a *compressive* force, can be obtained. They allow for factor of safety and for the factor for live load, so that the actual load applied should be taken. The use of the tables is explained by the following examples:—

Ex. 1. To find the size of a round spar to act as a trestle leg 15 feet from transom to ground, the load transmitted from transom to leg being 2 tons. On Plate 57 the horizontal line representing an unsupported height of 15 feet meets the 2 tons curve just at the vertical line indicating the use of a spar 8 inches mean diameter.

In the case of derricks, &c., the compression in the various members due to a weight w are as follows, under the most unfavourable circumstances in actual practice :—

Single derrick	...	1·5w	Sheers, leg with leading	
Swinging derrick—			block	·9w
Standing arm	...	1·7w	Sheers, other leg	·7w
Swinging arm	...	1·0w	Gyn, leg with leading block	·6w
Back strut	...	3w	do. other legs	·4w

Ex. 2. To find size of sheer legs 85 feet high to crutch, to lift a weight of 2 tons. Compression in leg with leading block is 9×2 tons $= 1·8$ tons. Following down between the curves representing 80 cwt. and 2 tons, it is seen that the horizontal line representing an unsupported height of 35 feet is met at a point that indicates the use of a spar nearly 12 inches mean diameter. Compression in the other leg is ·7 × 2 tons=1·4 tons, and, in a similar manner, the mean diameter of this leg is found to be 11 inches.

97. FORMULÆ FOR CALCULATING BUOYANCY.

1. In using closed vessels such as casks for floating piers, the *safe* buoyancy for bridging purposes may be taken at $\frac{9}{10}$ the *actual* buoyancy.

2. When calculating the buoyancy required for a raft or pier for a floating bridge, it is necessary to work to the actual weight of the load to be carried, *plus that of the superstructure*. The superstructure of the actual roadway of a bridge of normal width to carry infantry in fours may be taken at 120 lbs. per foot run, up to 15 feet span.

3. The buoyancy of closed vessels can be determined with sufficient accuracy by the following methods:—

(*a*) When the contents are known—

Multiply the contents, in gallons, by 9, this will give the safe buoyancy in pounds.

(*b*) For casks, when the contents are not known—

Actual buoyancy $= 5C^2L - W$ lbs.

Safe buoyancy $= \frac{9}{10}\{5C^2L - W\}$ lbs.

Where C is the circumference of the cask, in FEET, halfway between the bung and the extreme end; L is the extreme length, exclusive of projections, along the curve, in FEET; W is the weight of the cask in pounds.

4. The following are the dimensions, weight and buoyancy of certain casks:—

	Name of cask	Gallons.	Bung diameter.	Length along the cask L.	Circumference at ¼ length. C.	Weight empty. W.	Actual buoyancy. $5C^2L - W$	Safe buoyancy. $\frac{9}{10}(5C^2L - W.)$
			ins	ft	ft	lbs	lbs.	lbs
Used in the trades	leager	170	38·5	4·52	9·33	252	1,736	1,562
	butt	108	33·3	3·97	8·09	174	1,125	1,012
	puncheon ...	72	30·7	3·20	7·57	140	777	699
	hogshead ..	54	28·6	2·76	7·05	119	567	510
	barrel	36	25·3	2·42	6·23	88	382	343
	half hogshead ..	26	22·7	2·12	5·61	65	269	242
	kilderkin ...	18	20·3	1·81	5·02	49	185	166
	small cask ...	14	18·3	1·76	4·49	32	146	131
	,, ...	6	18·8	1·37	3·40	20	60	54

5. The buoyancy of a log can be obtained by multiplying its cubic content by the difference between its weight per cubic foot and that of a cubic foot of water. One cubic foot of water = 6¼ gallons, and one gallon weighs 10 lbs.

As, however, timber absorbs a great deal of water, only ⅚ of the *actual* buoyancy thus found can be relied upon.

Thus, the *safe* buoyancy of a pine log of which the cubic content is 96 cubic feet would be:—

$$\frac{5}{6} \times 96 \times (62\tfrac{1}{2} - 40)$$
$$= 80 \times 22\tfrac{1}{2}$$
$$= 1,800 \text{ lbs.}$$

40 lbs. being the weight of a cubic foot of pine.

The contents in cubic feet of an unsquared log of timber can be found by the following rule:—

$$\frac{L}{4} (D^2 + Dd + d^2).$$

Where L = length of log in feet,

D. *d* = diameter at ends in feet.

CHAPTER XIV.—HASTY DEMOLITIONS WITH EXPLOSIVES.

(See also Military Engineering, Pt. IV, Secs. 7, 10, 11 and 12.)

(For Stores, see Appendix III, Table 7.)

98. GENERAL REMARKS.

The service explosives available for hasty demolitions in the field are guncotton and cordite. Other explosives may sometimes be obtained, the most likely being gunpowder and dynamite. Where a lifting effect is desired, gunpowder should be used ; but, when a cutting or shattering effect is necessary, one of the others (high explosives) is better.

99. GUNCOTTON.

1. Guncotton is carried in the field for demolition purposes (*see* Appendix I). It is compact in form, safer than most others to transport and handle, and, like all high explosives, can be used without tamping.

Dry guncotton in small quantities will only burn away fiercely when ignited, and will not detonate if struck by a bullet. It will detonate if struck between two hard substances, and becomes more sensitive to percussion when heated through friction or by the sun.

If a small quantity of detonating substance such as fulminate of mercury be exploded in contact with dry guncotton, it will detonate with great violence, and also cause the complete detonation of any wet or dry guncotton with which it is in contact.

Guncotton will absorb about 30 per cent. of its weight of water, and when wet does not ignite or detonate easily, though the explosive force is slightly greater than when dry.

Consequently the bulk of guncotton is carried and used *wet* in the shape of "*slabs.*" For detonating this, *dry* guncotton, in the shape of small discs called "*primers,*" is also carried in air-tight tin cylinders, each containing 10 primers.

2. The slabs and primers for field service are as follows :—*

  i. Slab.—Weight 15 ozs. Dimensions 6 inches × 3 inches × 1⅜ inches, with one perforation for the primer. Each slab is in an hermetically sealed copper-tinned case.

  ii. Primer.—Weight 1 oz. Dimensions 1·35 inches to 1·15 inches in diameter by 1 25 inches long, with one perforation for the detonator. The primer is conical in form.

---

* The following may still be found, in addition to sizes used in the Naval Service :—
  (*a*) Slabs, 1¼ lb. and 1½ lb. (known as " S " and " T " respectively).
  (*b*) Primers, 2 oz. and 1 oz. (known as " F " and " H " respectively).
  These are cylindrical.

The slabs can be cut without danger with a sharp knife or saw, care being taken to press the guncotton between boards whilst it is being cut to prevent it flaking away. There is a special clamp in the R.E. equipment for doing this. The guncotton must be well soaked with water during the process.

3. A charge of wet guncotton is detonated by means of the explosion of a dry primer in close contact with it. The primer is exploded by means of a " *detonator*," containing an explosive which will detonate by the direct application of flame. The detonator is fired by means of either " *safety* " or " *instantaneous fuze*," which is lit by a fusee or other means.

If dry primers are not available, a piece of wet guncotton can be dried by exposure to the sun, and used instead. To test whether guncotton is dry, hold a piece of warm glass against it; if damp, a film of moisture will be left on the glass. When used in such places as bore holes in rock, guncotton must be dry; but a plastic explosive is more suited to such work.

4. A charge is connected up for detonation as follows :—

The fuze (safety alone or safety with instantaneous) is cut to the required length. (Sec. 104.) The end to be ignited is cut on a slant to expose as much of the composition as possible. The end to be inserted in the detonator is cut straight across. The straight cut end is then gently inserted into the open end of No. 8 detonator, from which the paper cap has been torn. This end of the detonator is then pinched (or with old-pattern detonator slightly bent) to make it grip on the fuze and so prevent its being withdrawn.

(Cavalry pioneers carry detonators with two feet of safety fuze ready fixed, the fuze having a piece of quickmatch added to the end to facilitate lighting.)

The primer having been tested to receive a detonator is placed in close contact with one of the slabs of the charge, either in one of the holes or tied to a slab, and the small end of the detonator is gently inserted into it so as to fill the entire length of the hole. If the hole is too large, a piece of paper or grass must be wrapped round the detonator to make it fit tightly; if too small, it must be enlarged with a rectifier or piece of wood, *but not with the detonator*.

5. THE CHARGE MUST BE IN CLOSE CONTACT WITH THE OBJECT TO BE DEMOLISHED AND EACH SLAB MUST BE IN CONTACT WITH THOSE NEXT IT. (Pl. 61, Figs. 1 and 2.)

The charge must extend across the whole length of the object to be cut. One detonator is sufficient for a continuous guncotton charge.

Arrangements must be made to prevent sparks from the fuze falling on the guncotton, and so setting it alight instead of detonating it.

6. The amount of guncotton (untamped) required for various charges can be calculated or obtained direct from the following table.

GUNCOTTON.

Object attacked.	lbs.	Remarks.
Masonry arch—haunch or crown.	$\frac{3}{4}BT^2$	Continuous charges.
Masonry wall—up to 2 ft. thick.	2 per foot	Length of breach B not to be less than the height of the wall to be brought down.
Masonry wall—over 2 ft. thick.	$\frac{1}{2}BT^2$	
Masonry pier	$\frac{3}{4}BT^2$	
Hard wood—stockade or single.	$3BT^2$	In a single charge outside. For a round timber charge $= 3T^6$
Hard wood—auger hole	$\frac{3}{4}T^2$	Where the timber is not round, T = smaller axis
Stockade of earth between timber up to 3 ft 6 ins. thick.	4 per foot	(Single charge outside.
Heavy rail stockade ..	7 per foot	
Fort gate	50	
Breechloading guns ..	—	For 3-inch gun use 2 lbs. Double the charge for every inch increase in calibre.
First class rail	1	Charge fastened against the web near a chair (if used).
Iron or steel plate ..	$\frac{3}{8}Bt^2$	t is in INCHES.
Frontier tower, stone and mud.	—	5 lbs. plus 1 lb. per foot of longer side if rectangular, or of diameter if circular. In one charge in centre of tower.
Steel wire cable	1	Up to 5 inches circumference; above 5 inches $\frac{C^2}{24}$ C being circumference in inches.

(Note in Remarks column, written vertically beside the Hard wood rows: "Soft wood half this")

Where B = length to be demolished in FEET.

T = thickness to be demolished in FEET.

t = thickness to be demolished in INCHES (in the case of steel or iron plate only).

NOTE.—The charge is in lbs; if the charge is tamped, the amount can be halved. Masonry includes concrete, stone, or brickwork.

In the presence of the enemy, charges may be placed hurriedly, and so under unfavourable conditions, and should therefore be increased by 50 per cent.

For emergency purposes BT^2 is effective with all classes of masonry and $2Bt^2$ for all steelwork. A slab will cut its own thickness of steel plate.

100. GUNPOWDER.

1. Gunpowder is not so useful for most hasty demolitions as gun-cotton, except where a lifting and shaking effect is required, and the larger the grain the less suitable is it, owing to its slow rate of burning.

2. Gunpowder charges must be tamped, and should be made up in as compact a form as possible. The powder should be placed in a well-tarred sandbag, or, failing that, in one sandbag inside a second one. About half the powder should first be poured into the bag, and then the safety fuze, knotted round a stick to prevent its being pulled out, should be inserted, a piece of stout wire or a withe being also attached to the stick, to help to support the fuze after it leaves the mouth of the bag. The rest of the powder is then poured into the bag, and the mouth secured with spun yarn (Pl. 61, Figs. 3 and 4), so as to make it easier to carry, a last seizing of the spun yarn being made round the fuze so that any pull on it will fall on the spun yarn and not on the fuze itself.

THE LENGTH OF FUZE TO BE EMPLOYED SHOULD BE MEASURED OUT-SIDE THE BAG.

3. A service sandbag will hold about 40 lbs. of gunpowder, which is about as much as a man can carry conveniently. When a charge has to be placed under fire, and the amount is greater than this, it should be divided amongst several bags, rather than put into one large one. In any case only one bag need be fuzed.

4. A gunpowder charge should generally be divided up into equal portions, which should be placed at a distance apart equal to twice the thickness of the object to be demolished. The several portions must be fired simultaneously. (Pl. 59, Fig. 7.)

5. The amount of gunpowder (tamped), required for various charges can be calculated from the following table.

GUNPOWDER.

Object attacked.	lbs.	Remarks.
Masonry arch—haunch or crown .. .. Masonry wall	$\frac{3}{7}BT^2$	Total amount divided into charges and placed at intervals of about twice the thickness of the masonry.
Wood stockade—hard wood.	40 to 100	One charge. Soft wood half this.
Stockade of earth between timber up to 3ft 6in. thick.	100 per 7 ft	Charges twice the thickness of stockade apart.
Fort Gate	200	One charge.
Tunnels	$\frac{1}{3}T^3$	Where T = total distance from the surface of the lining to the charge.

Where B = length to be demolished in FEET
T = thickness to be demolished in FEET.

NOTE.—In the presence of the enemy charges may be placed hurriedly and so under unfavourable conditions and should therefore be increased by 50 per cent.

101. CORDITE.

1. Cordite for demolitions may be obtained from gun cartridges.

For demolition purposes cordite is uncertain in its action and should only be used when no better explosive is available. The small sizes are the more reliable.

It is fired similarly to guncotton ; a primer of guncotton or other high explosive must always be used.

2. A cordite charge should be bound up tightly and fixed firmly against the object of demolition. It should in most cases be tamped, but this is not necessary when used against metallic structures, e.g., rails and girders. No portion of a cordite charge should be more than 12 inches from a primer, since the rate of communicating detonation is slow.

An untamped charge must be well protected from possible sparks from the fuze, since cordite ignites very easily.

3. Where good contact can be obtained, the power of cordite may be taken as equal to guncotton, otherwise a cordite charge should be increased by about 25 per cent.

102. DYNAMITES.

1. Dynamite, gelignite, gelatine dynamite, and blasting gelatine are often procurable in a mining locality, and can be used instead of gun-cotton. For military purposes the only advantage they have over guncotton is that, being plastic, they are easier to fit into narrow and irregular holes and spaces as for blasting rock and cutting steel work. For demolishing masonry dynamite it is not so good as gun-cotton, as its action is even more local.

It cannot be used after exposure to wet, which separates the nitro-glycerine and makes it dangerous.

2. Dynamites freeze at 40° F., and remain frozen at higher temperatures. When cold weather is likely, they should be buried a foot or two underground. Frozen dynamites can be distinguished by being harder than unfrozen, by being more brittle than plastic, and being of a slightly lighter colour. Frozen dynamites should be thawed before use, but this must be done with great care and only small quantities should be dealt with at a time.

FROZEN DYNAMITES MUST NOT BE WARMED ON OR NEAR FIRES, STOVES, OVENS, OR STEAM-PIPES; NOR EXPOSED TO THE DIRECT RAYS OF A TROPICAL SUN. THEY CAN BE THAWED IN A WATER-TIGHT TIN CAN, WHICH SHOULD BE PLACED IN A VESSEL CONTAINING WATER PREVIOUSLY AND SEPARATELY HEATED TO A TEMPERATURE NOT HOTTER THAN THE WRIST CAN BEAR. A PROPER ' DYNAMITE WARMING PAN,' WHICH CANNOT BE PLACED ON A FIRE WITHOUT DESTROYING IT, SHOULD ALWAYS BE USED IF POSSIBLE.

3. They are usually obtained in 2 oz. cartridges wrapped in parch-ment paper, in boxes of 5 and 50 lbs.

4. They can be detonated by fuze and No. 8 detonator, or by fuze and cap; the former is unnecessarily strong.

They are fired in a similar manner to guncotton, but no primer is required. The detonator with fuze attached, is inserted into a hole made in one of the cartridges, and tied in position. This hole must be made with the rectifier or a piece of wood. Where a dynamite charge requires ramming, as in a bore hole, each cartridge must be gently squeezed into place with a wooden rammer, the fuzed one being the last.

5. The relative strengths of the common explosives, when well tamped, may be taken as follows:—Gunpowder 5, Cordite 8, Dynamite 9,

Guncotton 10, Gelignite 10, Gelatine Dynamite 12, and Blasting Gelatine 18. For use under water, dynamite and similar explosives should be tied in a waterproof bag.

6. Although complete detonation is impossible with frozen dynamite, an explosive effect roughly half that of thawed dynamite can be obtained by using a thawed cartridge as primer. Frozen dynamite will explode by simple ignition at a temperature of about 360° F.

103. DETONATORS.

1. Pl. 59, Fig. 1, gives a section of the service No. 8 detonator. It consists of a brass tube painted red, the small end of which, A, contains the detonating compound (fulminate of mercury); above this is a wooden plug with a hole in it, through which passes a piece of quickmatch. The upper end of the tube is empty, for the insertion of the fuze, and is closed by a small paper cap. These detonators are packed in tin sealed cylinders painted red, which contain 25. No. 8 detonator will detonate dry guncotton, but it will *not* detonate wet guncotton or cordite without a primer. Both safety and instantaneous fuze can be connected to it. The older Mark III detonator may still be met, it is longer than Mark IV and has no lugs. A Mark V detonator has now been introduced. It is similar in appearance to Mark IV, but the body is made of copper.

2. Where dynamite, &c., is found, a smaller kind of detonator used in civil works will often be available and should be used in preference to No. 8 detonators as being more economical. This is made of copper, and contains less fulminate than the No. 8 service detonator. (Pl. 59, Fig. 6.)

These "caps" vary in size and strength (Sizes 3 to 10). To detonate dynamite, size 8 are used as a rule. The weaker sorts cannot be counted on to detonate guncotton primers, but for all dynamites, Size 6, 7 or 8 (the latter for blasting gelatine), are strong enough.

They can be connected up to safety or instantaneous fuze.

3. Detonators must be stored apart from explosives. To prevent them being exploded by stray bullets, they may be buried.

104. SAFETY FUZE.

1. The present pattern of safety fuze is known as "Safety, No. 9." It consists of a train of fine gunpowder enclosed in flax twist, covered with guttapercha and waterproof tape. It is packed in tin cylinders containing 8, 24, or 50 fathoms.

It is coloured BLACK.

Safety fuze will burn under water.

2. For practical work the rate of burning can be taken at 4 feet per minute; but all fuze should have its rate of burning tested before being used. Fuze which has been more than six months or so in a tropical climate should be very carefully examined and tested.

3. It is difficult to light safety fuze with a match or flame. A portfire or vesuvian (fusee) is best, but, in the absence of such means of ignition, the head of a match inserted in the fuze and lit by another match or rubbed by the prepared portion of a matchbox, is usually successful.

105. INSTANTANEOUS FUZE.

1. Instantaneous fuze consists of three strands of quick match enclosed in flax and several layers of guttapercha and waterproof tape. It is

8609

coloured ORANGE. It is used, in the absence of electric appliances, when the time taken to burn the requisite length of safety fuze would be inconveniently long.

2. It burns at the rate of 30 yards a second, or practically instantaneously; it is packed in sealed tins holding 100 yards.

3. It can be distinguished in the dark from safety fuze by feeling the open crossed thread, snaking outside.

Unless the safety of the firer or the depth of the tamping requires it, instantaneous fuze should not be used in addition to safety fuze where there is only one charge, as it increases the liability to missfire.

4. It is liable to jerk the detonator out of a charge by its violent jump on ignition, unless it is carefully secured near the detonator.

106. Substitutes for Fuzes.

1. If the service fuzes are not available; gunpowder, mixed with water, ground into a fine paste between two pieces of wood, and then pressed into a tube, may be used instead of safety fuze. Its rate of burning will vary according to the dampness of the powder.

2. Instead of instantaneous fuze, powder hose made by filling tubes of linen with fine powder may be used. The tubes may be from ½ to 1 inch in diameter, and up to 20 feet in length. The rate of burning will vary from 10 to 20 feet per second.

107. Firing Charges.

1. When firing charges with instantaneous fuze, a piece of safety fuze should be joined on for lighting, in order to allow time for getting away, except as described in para. 5 below.

2. Safety and instantaneous fuze may be joined in three ways :—

    (a) Cut the instantaneous fuze on the slant so as to expose the quickmatch for a short length, and treat the safety fuze in the same way, taking care that the composition is well laid open. Join these two surfaces together and bind up tight. A small piece of wood is useful as a splint, and, if available, a little powder or quickmatch can be put between the two fuzes. (Pl. 59, Fig. 2.)

    (b) Place the ends of both fuzes in a small bag of gunpowder, exposing the quickmatch of the instantaneous fuze for two inches.

    (c) Cut a semi-circular nick in each fuze near the ends, exposing the core of each. Put the nicks together as shown in Pl. 59, Fig. 5, and tie them together tightly.

3. To join two lengths of instantaneous fuze, slit the outer covering of each piece of instantaneous fuze at the end, so that it can be turned back to expose the quickmatch; the strands are then twisted together, the outer covering made to overlap the joint, and firmly fixed with twine.

4. Joints in fuze can be made waterproof by wrapping them round with indiarubber tape smeared with indiarubber solution, which are articles of R.E. equipment. Ordinary tape and tallow would do for a short time against damp.

5. When several charges are to be fired simultaneously and electricity is not available, it may be done as shown in Pl. 59, Fig. 7, by using equal lengths of instantaneous fuze, "bc, bc_1, bc_2," which are ignited at "b" by a length of safety fuze, "ab."

The joint at "*b*" can be made with a small bag or box of gunpowder or flaked guncotton, into which the end of the piece of safety fuze and the ends of the instantaneous fuze are led, the quickmatch in the latter being exposed for two inches. Care must be taken that the *lengths of instantaneous fuze are equal*, irrespective of the distance from the powder box to the charges. This method cannot be relied on to give certain results.

108. Precautions to be Observed in Firing Charges.

1. When a demolition is to be carried out under fire every precaution should be taken against a possible failure; spare men should be detailed to replace casualties amongst those carrying the stores and every man with the party should have the means of lighting the charge, and should know exactly what is to be done and the means available.

2. For large charges of all sorts which cannot easily be got at after tamping, and for demolition work where certainty and rapidity are essential, it is a good rule to insert and light two fuzes (and detonators if required) in the charge in case one should prove faulty.

3. When pinching or bending the mouth of a detonator or cap to grip the fuze, care should be taken not to squeeze the detonating end, or to point it at anybody.

4. When tamping a guncotton charge with earth, stones, &c., the detonator should be protected from blows, and the fuze from wrenches which might displace it. The fuze should also be lightly fixed or weighted so that, when lit, it will not curl up and set the charge on fire.

5. When a charge fails to explode, it should not be approached for at least half an hour. In accessible places, the charge should be "killed" by detonating a fresh charge as close as possible to it. When it is necessary to withdraw the charge, the tamping must be carefully cleared away from the detonators with pieces of wood, the whole being previously drenched with water, and the detonators removed at the earliest opportunity.

6. The person who orders a charge to be fired is responsible that surplus explosives and detonators, &c., have been removed to a place of safety, and that steps have been taken to ensure the safety of people ignorant of the place and time of the explosion.

7. Where several charges are to be fired without awaiting successive explosions, all fuzes within the danger area of any one charge should be lit by one person only, allowance being made in the length of the fuzes. They should not be lit by men acting independently.

109. Brickwork and Masonry Demolitions.

1. For the demolition of brickwork or masonry with guncotton, the charges worked out by the formulæ in the table will sometimes be too small to allow the whole of the length to be cut, to be covered with slabs in contact with one another. In this case add extra slabs. The old pattern slabs, which are larger, may be halved.

2. When it is desired to make a gap in a bridge or viaduct in which there is a series of arches, the best result is got by cutting a pier and so bringing down two or more arches for each charge, but in hasty demolitions this can only be done when the piers are thin.

The best explosive to use for this purpose is guncotton. The charge should be placed where the section of the pier is smallest, and if possible a groove should be cut in the pier for the charge; this reduces "T" and also to some extent tamps the charge. Otherwise the charge should be tied in a continuous strip along a board, which is then fixed with the guncotton next the pier (Pl. 60, Fig. 1).

If gunpowder is to be used, the whole charge should be divided up as described in Sec. 100 (4), and placed in pits dug at the foot of the pier to be destroyed. If the piers are in water or if time is important the arches should be attacked.

3. The amount of guncotton and gunpowder required for cutting short and thick piers is prohibitive, and the arches should be attacked. The best method of doing this is to place a charge at *each* "haunch" of the arch (Pl. 66, Fig. 5). This ensures a much larger gap being made than if only one charge were placed at the "crown." (Fig. 2.)

If guncotton is used, a trench must be dug down to the back of the arch ring at each haunch, then the slabs should be laid along the trench on the back of the arch ring. The amounts of explosive required to cut the portions of arch ring protected by the parapet walls being piled up, at the ends of the trench, against those walls. Tamping may be used, but is not essential.

If gunpowder be employed, the charge for each haunch should be divided as in Sec. 100 (4), the outside ones being placed twice the thickness of the arch-ring from the side walls, to avoid the charges blowing out through the latter. A pit must be dug for each portion of the charge down to the back of the arch, and tamping is necessary equal to twice the thickness of the arch.

In all cases the charges at *both* haunches should be fired simultaneously. Sec. 107 (5).

4. When there is not enough time to reach the haunches, the crown may be attacked in a similar way, but the result is not so satisfactory (Pl. 60, Fig. 5).

5. Where the filling over the crown is deep or traffic must not be interrupted, arches can be cut by guncotton without digging through the roadway. The charge to cut through the arch at the crown is tied in a continuous strip along a plank, and held up underneath the arch by ropes from the parapet, with the guncotton next the arch. These ropes should be windlassed up tight so as to ensure contact between the guncotton and the arch; and the plank should be supported or trussed to prevent sagging in the middle (Pl. 60, Fig. 2).

If the demolition of a bridge does not immediately follow the preparations, steps must be taken to preserve fuzes, &c., from injury by traffic or weather.

6. To demolish a wall by guncotton, a groove should be cut for the charge in the wall; if this is not possible, the charge must be laid against the wall. With gunpowder the total charge should be divided up, as in Sec. 100 (4), and earth tamping must be used (Pl. 60, Figs. 3 and 4.)

To bring down the top of a wall, the charge must extend over a length not less than the height of the wall.

7. For weakly-built houses, place a charge in the centre of each room, shutting all doors and windows. In houses with chimney stacks the charges should be placed on the hearths of fire-places. If possible, fire charges simultaneously by electricity. The amount of explosive required depends on the size of the rooms and the nature of the walls. Mud huts up to 18 ft. square, with walls 2 feet thick at the bottom, have been destroyed by about 4 lbs. of guncotton placed inside the hut in one corner, all openings

being closed; 6 to 12 lbs. of guncotton will probably destroy a four-roomed cottage. For strongly-built buildings it may be necessary to attack the walls.

8. Towers such as those in the North-west frontier of India are generally hollow or with a solid base up to about 15 feet high. The walls are usually 3 to 4 feet thick, and the towers circular or square in plan. For hasty demolition, place the charge in a hole 2 or 3 feet deep in the centre of the floor and tamp it well.

The charge of guncotton may be obtained by adding 5 to the inside length of the diameter or side, measured in feet, and allowing 1 lb. for every foot in the total.

110. TIMBER DEMOLITIONS.

1. It is more economical to destroy baulks of timber by cutting them down or burning them than by explosives, which should only be used when time presses.

Guncotton or dynamite is the best explosive for this purpose and is most economically used when placed in auger holes bored horizontally at the required height. For baulks up to 18 inches diameter one auger hole bored right through, if necessary to contain the charge, will suffice; the middle of the charge coinciding with the centre of the timber. For larger baulks two or more holes bored alongside each other will be needed.

If of guncotton the whole charge should consist of primers, earth or clay being used for tamping. The fuze may be hung on a nail or splinter to take the weight off the detonator.

Timber may be made to fall in any required direction by getting a strain on it beforehand with a rope.

2. The most convenient way to prepare a guncotton charge for use against a stockade, wall, &c., is to tie the slabs beforehand on to a board so as to ensure their being in contact with one another; a hole may have to be cut in the board for the detonator and fuze (Pl. 61, Figs. 1 and 2). The board can then be carried up, placed with the guncotton between it and the stockade, and two pickets driven into the ground to keep it in position, or it may be hung on a couple of nails hammered into the stockade.

3. Before attempting to place a gunpowder charge against a gate or stockade, the men carrying the charges and tamping bags should be thoroughly drilled as to how the charge and tamping is to be placed. The men carrying the powder bags on their shoulder lead the way, place the bags against the stockade, fuze down-wind, so that sparks will not blow into the charge, and one prepares to light. The other men, each carrying a bag of sand, place them as shown in Pl. 61, Fig 5. The fuze is then lighted, and all get under cover as quickly as possible.

4. The gate of a fort may be treated as a very strong stockade. As the thickness cannot usually be known, a good margin in the amount of the charge should be allowed. When guncotton is used 50 lbs. will usually be enough. The charge may be either placed on the ground or hung to the gate on a nail. When gunpowder is used a charge of 200 lbs., tamped with sandbags, should suffice.

111. RAILWAY DEMOLITIONS.

1. On railways, the easiest parts to attack in hasty demolitions are the bridges.

Where there is a choice between masonry and iron girder bridges, the girder bridge ought, as a rule, to be attacked, as its demolition will be quicker, guncotton will be economized, and the result will usually be greater.

Masonry arch bridges should be attacked as described in Sec. 109.

2. Iron and steel bridges *may* be destroyed by placing charges of gunpowder or guncotton beneath the ends at the supports; but the quickest way with small girders is to actually cut the girders themselves with guncotton.

The best position in which to place the charge varies in different kinds of bridges, *e.g.* :—

(*a*) When the girder is of uniform section and only extends over one span, it is near one point of support (Pl. 62, Fig. 1). This form of girder is met with in small bridges only.

(*b*) When the girder depends for its strength principally upon a lower tension boom, cut the lower boom for certain (Pl. 62, Fig. 2).

(*c*) When the girder is of the type shown in Pl. 62, Fig. 3, cut the lower boom, and if possible the strut Z and top boom. Sometimes a charge at Y (Fig. 3) can be used to cut out the end bearings from under the girders, let down both adjacent girders and shatter the top of the pier.

(*d*) In a plate girder not of uniform section, only extending over one span, it is just before the first thickening plate on the flanges. (Pl. 62, Fig. 4.) If continuous over several spans it is on the shore side of the first pier.

All girder bridges have at least two main girders which carry the flooring and go right across the span, and these main girders alone need be attacked.

3. Nearly all girders consist of a top and bottom "*flange*" or "*boom*," connected by a "*web*," which may either consist of continuous plating or of open cross bracing.

Pl. 64, Fig. 1 represents a typical section of a girder and the arrangement of the charges upon it will serve as a guide for most occasions. The amount of guncotton required for its demolition is obtained as follows :—

(*a*) Flanges. Taking the formula for steel plate, the breadth of each flange will represent " B," while " t " will be the thickness of the flange plates, plus the thickness of the angle-iron (if any), plus the height of one rivet head (if any).

(*b*) Web. This is considered as consisting of 3 parts. The thicknesses of the top and bottom portions are taken as the thickness of the single plate forming the main web, plus the maximum thicknesses of the angle irons (if any), plus the height of one rivet head. The breadths " B " are measured from the angle of the angle-iron to its edge along the web. The centre portion, between the top and bottom angle-irons, is taken as a single plate.

The total charge so obtained is divided into two parts, in proportion to the metal in the upper and lower halves of the girder (often equal), and placed in the diagonally opposite angles formed by the junction of the web with the top and bottom flanges. (Pl. 64, Fig. 1.) The spaces between the rivet heads must be packed with clay to provide a uniform surface for the charge to rest on. The latter is kept in position by strutting it against the opposite flange. The two charges must be fired simultaneously.

4. The charge for the demolition of the girder shown in Pl. 64, Fig. 1 is worked out as follows, using the ordinary formula $\frac{2}{3}$ Bt^2:—

Top flange $\frac{2}{3} \times 1\frac{1}{4}$ (breadth) $\times \left\{ \frac{1}{2}'' \text{ (thickness of flange)} + \frac{1}{2}'' \right.$ (thickness of angle iron) $+ \frac{1}{4}''$ (height of one rivet head) $\left. \right\}^2$

$= \frac{2}{3} \times \frac{5}{4} \times (\frac{5}{8})^2 = 4\cdot22$ lbs.

Bottom flange $\frac{2}{3} \times 1\frac{1}{4}'$ (breadth) $\times \left\{ \frac{1}{4}'' \text{ (thickness of flange)} + \frac{1}{8}'' \right.$ (thickness of angle iron) $+ \frac{1}{4}''$ (height of one rivet head) $\left. \right\}^2$

$= \frac{2}{3} \times \frac{5}{4} \times (1\frac{5}{8})^2 = 4\cdot95$ lbs.

Web, top portion $\frac{2}{3} \times \frac{5}{12}$ (breadth) $\times \left\{ \frac{3}{8}'' \text{ (thickness of web)} + \right.$ $2 \times \frac{1}{2}''$ (two thicknesses of angle iron) $+ \frac{1}{4}''$ (height of one rivet head) $\left. \right\}^2 = \frac{2}{3} \times \frac{8}{12} \times (1\frac{7}{8})^2 = 1\cdot82$ lbs.

Web, bottom portion $\frac{2}{3} \times \frac{3\frac{3}{4}}{12}'$ (breadth) $\times \left\{ \frac{3}{8}'' \text{ (thickness of web)} + \right.$ $2 \times \frac{1}{8}''$ (two thicknesses of angle iron) $+ \frac{1}{4}''$ (height of one rivet head) $\left. \right\}^2 = \frac{2}{3} \times \frac{3\frac{3}{4}}{12} \times (1\frac{7}{8})^2 = 2\cdot19$ lbs.

Web, central portion $\frac{2}{3} \times \left\{ 8' - (2 \times \frac{1}{12}' + \frac{4\frac{1}{2}}{12}' + \frac{3\frac{3}{4}}{12}') \right\}$ (breadth) $\times \left\{ \frac{3}{8}'' \right\}^2$ (thickness of web) $= \frac{2}{3} \times 2\frac{1}{4} \times (\frac{3}{8})^2 = \cdot47$ lbs.

The total charge is 13·15 lbs. or just 14 slabs. This should be placed as shown in the figure in two charges of 6 and 8 slabs respectively.

So long as there is sufficient explosive available, however, one of these charges should be increased so as to cover the web between the charges.

5. For hasty demolition of railway girder bridges when there is not time to measure the section of the girder, the following formula will give sufficiently accurate results within the limits stated:—

$$C = \frac{L^2}{15D}$$

where C = charge of guncotton in slabs (15 oz.), including the allowance of 50 per cent. for the presence of the enemy.

L = length of girder in feet.

D = total depth of girder in feet.

This formula gives the charge required for one girder of *single line* of standard railway. The charge must be placed near an abutment and be divided up and fixed to the girder in the manner described above.

The formula is applicable to girders of spans varying from 20 feet to 80 feet. Where one girder has to bear the whole load of a line of railway, *e.g.*, two girders carrying a double line, or a centre girder carrying half of two single lines, the amount given by the formula should be doubled. This will, however, give rather more than the necessary charge.

Example:—The girder shown on Plate 64, Fig. 1, requires a charge of 14 slabs, or, allowing for the 50 per cent., 21 slabs. The above formula gives 20 slabs.

6. To cut a rail the charge should be tied tight into the web close to a chair on the same side as the key (Pl. 64, Fig. 2), or in the case of a heavy unchaired rail, as Pl. 63, Fig. 3. In the hasty demolition of a railway line care must be taken that the break is sufficiently broad, or sufficient damage is done, to ensure the stopping of traffic.

7. An effective way of damaging a railway line is by firing charges at points and crossings, in the positions shown in Pl. 63, Figs. 1 and 2.

8. In carrying out the destruction of girders and rails with high explosives, it should be recollected that fragments are liable to be blown 1,000 yards away from the spot where the demolition is being carried out.

9. Blowing in tunnels is a good way of stopping traffic, but to be effective a considerable amount of time and a large quantity of explosive are required.

The points attacked should be some distance within the tunnels, and it is better to blow down one long tunnel in several places than several tunnels in one place only. The crown or the haunches should be attacked as in cutting arches, and the lining should be brought down for some distance along the length of the tunnel. In hard soil not much damage will be done by merely cutting the lining, as very little of the soil may fall. The charges should be placed in chambers, branching off the gallery dug in from the surface of the tunnel, as far back from the interior surface of the arch as time and explosive available will allow, and twice as far from each other as from the surface.

Gunpowder is the best explosive for this purpose.

For calculating the charge, T should be taken as the total distance from the surface of the lining to the charge.

112. Destruction of Guns.

To destroy a gun with guncotton (or cordite), a shell having been loaded in the ordinary way, the charge necessary for the destruction of the gun should be packed in behind it so as to be in close contact with the shell and with the sides of the chamber. After the insertion of the primer, sods, earth, paper or other material that may be at hand should be used to keep the guncotton in position. The breech should then be closed as far as possible, just allowing room for the safety fuze or electric leads for igniting the charge. A shell is not absolutely necessary for destroying a gun by this method, but, if available, its use increases the effect.

The charges required for guns from 3-inch to 6-inch calibre are given by the following rule :—

" For a 3-inch gun use 2 lbs., and double the charge for every inch increase in calibre, *e.g.*, for a 4-inch gun use 4 lbs., and for a 5-inch, 8 lbs. With cordite charges add 20 per cent."

Muzzle loading guns can be demolished by placing a guncotton charge at the bottom of the bore and tamping with sand or water; 1½ lbs. is sufficient for light, 4 lbs. for heavy guns up to 6-inch.

CHAPTER XV.—HASTY DEMOLITION OF RAILWAYS AND TELEGRAPHS WITHOUT EXPLOSIVES.

113. RAILWAYS.

1. The method in which a railway is attacked must depend largely on the time at the disposal of the working party, its numerical strength, and on the extent of damage which it is desired to carry out. The General Staff is responsible for issuing the necessary instructions as regards the amount of damage to be done. (*See also* F.S. Regs., Pt. I, Sec. 71.)

2. When a railway is to be interrupted the first step in every case is to sever or block the main lines of rails. The most important technical tools should also be removed as well as all individuals entrusted with the working of the railway; and the signals, first the electric and then the visual, should be destroyed. Of the permanent way, points and crossings are the most important parts, and should be destroyed or removed as soon as the preliminary severance of the through lines has been effected.

3. Buildings not being indispensable to the traffic are seldom worth destroying; but workshop fittings, telegraphic apparatus and batteries should be taken away, and handed over to the Director of Telegraphs, and stationary engines made unserviceable by taking out the piston, &c. The water supply of a line may also be attacked, and the more complete the destruction of tanks and pumps the more efficient will be the obstruction. All fuel, if not required for immediate use or removal, should be burnt.

4. If the rolling stock cannot be removed, it may be rendered unserviceable by burning; or trains may be run against each other at full speed on the same line, or over an embankment by turning a rail.

Locomotives may be rendered useless, but still repairable, by taking off the injector, the connecting rods from cross-heads to cranks, or the piston or safety valve.

In carriages the springs may be removed so as to let the body of the carriage fall on the wheels and axles.

5. A simple method of attacking the permanent way is to remove switches or portions of the line at intervals, especially at curves, &c., and carry them away. To remove the rails, the fish-plate nuts should be unscrewed with a spanner; but, if one is not available, they may generally be broken off by hammering. The enemy will find considerable difficulty in fitting in rails of the right length in the demolished portions, but, if this method is adopted on a double line, one complete line of rails and part of the other must be entirely removed, otherwise the enemy might employ the available material to form a single line for temporary service. It is better systematically to collect and remove the whole supply of one article essential to the working of the railway than to effect promiscuous but incomplete damage of several different things.

6. A second method, where many men are available, and where the time is short, is to attack the line at several points at once, tear up the permanent way and render it useless on the spot. Labourers can be employed in preparing sleepers in piles for burning, placing rails upon them and then twisting them. If the rails are only bent they can be straightened and used again, but if twisted they must be sent to regular workshops to be re-rolled before they can be utilized. The chairs should be broken by a sledge hammer.

7. A third method is to lift up and turn over whole portions of the railway, together with the attached sleepers. This method is specially useful on high embankments. Having disconnected the rails at both flanks and cleared the sleepers of ballast, the men are formed along a rail in single rank, outside of it and facing inwards. At a signal they seize the rail, lift it up with the sleepers attached, and turn it over. Teams of horses or oxen can be hooked on to the rails and used in like manner.

8. A fourth method is to divide the party into squads of ten men each, and to equip each squad with two iron hooks (Pl. 64, Fig. 3) and two ropes, each six yards long and two levers. The irons are then fixed as shown. One end of the rail having been disconnected, the ropes attached to the ends of the levers are hauled on, the rails twisted and the spikes drawn. Each rail requires about five minutes' work, so that in one hour a squad can destroy twelve lengths of rail.

9. A fifth method of demolition is to take up the permanent way and remove it bodily in wagons. This is the most satisfactory method, but requires much time and careful arrangement of the necessarily large working parties. (*See* Mil. Eng., Pt. IV, para. 296.)

114. TELEGRAPHS.

1. It is assumed that the line to be destroyed lies in a country occupied by the enemy, to which access has been obtained for a short time by a raiding party: otherwise it would of course be easy either to disconnect the wires and appropriate them, or, leaving the lines intact, to interpose instruments, and thereby read any messages sent by the enemy.

2. The amount of damage that can be done in a short time to a line of telegraph depends chiefly on the number of separate wires running parallel to each other on the same poles in the case of an aerial line, or the number of separate cables contained in the same set of pipes in a subterranean line. These forms are by far the most likely to be encountered on service.

3. Wooden poles can be readily cut down, the easiest and safest to attack being those that have stays.

A rope having first been fixed to the top of the pole or thrown over the wires, the pole should be partly cut through at about 4 feet from the ground. All hands should then commence to strain on the rope, except one man, who should cut the stay through with a file or pliers. The men on the rope must be sufficiently far from the pole to be well clear of the wires when they fall.

The destructive effect will be increased if, previous to this, the poles on each side are partly cut through and their stays destroyed.

Cast iron poles can easily be broken with a sledge hammer.

Having brought down as much as possible of the line in this way, the wires should be cut and twisted up so as to be rendered useless. The insulators should also be broken.

Any damage of this sort, however, can be quickly repaired by the enemy using cable, and even the complete restoration of the poles and wires will not take very long to accomplish.

4. If possession can be obtained of an office, wires can be disconnected. Any papers connected with the working of the line and, if possible, the instruments, should be sent to the officer in charge of the field telegraphs.

Records of messages should be sent at once to the headquarters of the force to which the party belongs.

In making a raid on a telegraph office still containing operators care must be taken to disconnect the lines radiating from the office before disturbing the operators.

5. A subterranean line is naturally more difficult to discover than an aërial one; for this reason among others they are now extensively employed in countries liable to invasion. In England they are rarely met with except in large towns, where overhead wires are dangerous.

The existence of such a line being known or suspected, marks indicating the position of test boxes should be searched for. These marks are usually about 100 yards apart and generally consist of iron covers or stones numbered in succession.

If the general position of a line is known a cross-trench can be dug at right angles to its probable direction, about 2 feet deep, and in this way the pipes may be discovered. These can then be dug up and destroyed according to the means and time available, the wire being pulled out and cut to pieces.

If possible the trench should be carefully filled in and all traces removed.

6. A subaqueous line is rarely employed except for crossing seas or big rivers, but in time of war one might be laid along the course of a river to connect towns on its bank, as was done at Paris in the Franco-German war.

If the existence of such a line is suspected, the river should be dragged with a grapnel, and if found as large a piece as possible should be cut out of the line and destroyed.

SUMMARY OF TOOLS AND EXPLOSIVES CARRIED

TOOLS.	Weight.	Length.	Cavalry Regiment.			Horse Artillery Brigade.			Field Artillery Brigade, 18-pr.		
			Wagon and Cart Equipment	Tools.	Total	Wagon and Cart Equipment	Tools	Total	Wagon and Cart Equipment	Tools.	Total.
(a) Entrenching.	lbs. ozs. (A)	ft. in.									
Shovels, G.S.	3 8	3 1	2	20	22	62	76	138	62	144	206
Spades	5 10	3 2	...	...	...	...	...	...	...	...	1
Axes, pick	8 0	3 0	1	14	15	21	48	69	31	73	104
(b) Cutting.											
Axes { felling	6 7	2 8	1	6	7	21	12	33	31	18	49
hand	2 3	1 4	...	6	6	...	...	...	...	...	...
Hooks { bill	1 13	1 3½	1	14	15	21	48	69	31	72	103
reaping ...	1 0	1 6½	...	3	3	...	40	40	...	60	60
Saws { cross-cut ...	6 7	5 0	...	...	...	...	...	...	...	...	...
folding ...	1 12½	...	...	3	3	...	...	...	...	...	...
hand, in case ...	5 6	2 2	...	3	3	...	36	36	...	54	54
(E) Wirecutters (in frogs)	1 4	...	...	54	54	...	62	62	...	62	62
(c) Miscellaneous.											
Crowbars	12 0	3 6	...	3	3	...	...	...	...	...	...
Sandbags	...	...	...	150	150	...	...	...	...	...	...
Mauls, G.S.	2 11	...	...	3	3	...	...	...	...	...	...
EXPLOSIVES.											
Guncotton { dry primers field, 1 oz.	...	...	...	240	240	...	...	...	...	...	...
wet, charges field, 15 oz.	...	...	...	96	96	...	...	...	...	...	...

NOTE.—These figures are correct for the date of issue of this Manual.
of the

Field Artillery Howitzer Brigade.			Horse Artillery Battery.			Field Artillery Battery, 18-pr			Field Artillery Howitzer Battery			Heavy Artillery Battery			Heavy Battery Ammunition Column		
Wagon and Cart Equipment	Tools	Total.	Wagon and Cart Equipment.	Tools	Total.	Wagon and Cart Equipment	Tools	Total	Wagon and Cart Equipment	Tools.	Total.	Wagon and Cart Equipment	Tools	Total.	Wagon and Cart Equipment	Tools.	Total.
26	72	98	...	86	86	...	86	86	...	30	30	4	16	20	14	8	22
...	...	1	...	...	...	...	...	...	...	...	...	...	...	1	...	...	...
18	87	50	...	18	18	...	18	18	...	15	15	2	9	11	7	4	11
18	12	25	...	6	6	...	6	6	...	6	6	2	...	2	7	...	7
...	...	...	...	...	...	...	...	...	...	...	...	...	...	...	...	...	...
18	86	49	...	18	18	...	18	18	...	15	15	2	8	10	7	4	11
...	24	24	...	...	...	...	...	...	...	...	...	...	6	6	...	...	14
...	...	...	...	...	...	...	...	...	...	...	...	...	...	...	...	...	...
...	...	...	...	...	...	...	...	...	...	...	...	...	...	...	...	...	...
...	24	24	...	12	12	...	12	12	...	9	9	...	8	8	...	...	4
...	62	62	...	20	20	...	20	20	...	20	20	...	20	20	...	6	6
...	...	...	...	...	...	...	...	...	...	...	...	...	4	4	...	...	1
...	...	...	...	...	...	...	...	...	...	...	...	...	...	...	...	...	...
...	...	...	...	...	...	...	...	...	...	...	...	...	8	8	...	...	1
...	...	...	...	...	...	...	...	...	...	...	...	...	...	...	...	...	...
...	...	...	...	...	...	...	...	...	...	...	...	...	...	...	...	...	...

They should be corrected periodically in accordance with the F.S. Manuals various arms.

SUMMARY OF TOOLS AND EXPLOSIVES CARRIED

TOOLS.	ENGINEERS									Infantry Battalion.		
	Field Troop			Field Company			Bridging Train.					
	Wagon and Cart Equipment	Tools	Total.	Wagon and Cart Equipment	Tools	Total	Wagon and Cart Equipment	Tools	Total.	Wagon and Cart Equipment	Tools	Total
(a) Entrenching.												
(A)Shovels G.S.	2	37	39	2	111	113	10	40	50	2	(B)224	226
Spades	...	7	7	...	19	19	...	40	40	...	...	...
Axes, pick	1	39	40	1	(C)107	108	5	40	45	1	(D)150	151
(E)Implement (patn. 1908)	...	...	...	...	...	...	...	...	...	...	936	936
(b) Cutting.												
Axes { felling	1	25	26	1	47	48	5	40	45	1	16	17
hand	...	14	14	...	28	28	...	...	...	...	8	8
Hooks { bill	1	18	19	1	39	40	5	40	45	1	42	43
reaping	...	6	6	...	10	10	...	8	8	...	20	20
Saws { cross-cut	...	2	2	...	4	4	...	...	...	...	...	...
folding	...	4	4	...	8	8	...	...	...	...	(E)32	32
hand, in case ...	...	16	16	...	27	27	...	4	4	...	...	...
Wirecutters (in frogs) ...	...	18	18	...	41	41	...	44	44	...	24	24
(c) Miscellaneous.												
Crowbars	...	6	6	...	8	8	...	...	...	...	8	8
Sandbags	...	264	264	...	852	852	...	...	...	...	30	30
Mauls, G.S.	...	3	3	...	5	5	...	40	40	...	...	...
EXPLOSIVES.												
Guncotton { dry primers, field, 1 oz.	...	480	480	...	720	720	...	...	...	...	...	...
wet, charges field, 15 oz.	...	280	280	...	560	560	...	...	...	...	...	...

NOTE.—These figures are correct for the date of issue of this Manual.

of the

Mounted Infantry Battalion			M I Company or Squadron Irish Horse			Headquarters Infantry Brigade			Remarks
Wagon and Cart Equipment	Tools	Total	Wagon and Cart Equipment	Tools	Total	Wagon and Cart Equipment	Tools	Total	
2	240	242	...	6	6	2	120	122	(A).—The weight and length given are those of G.S. shovel. The R.E. shovel weighs 5 lbs. and is 8 feet 4 inches long.
...	...	...	...	...	...	...	...	...	
1	124	125	...	4	4	1	80	81	(B).—14 shovels are carried on each company pack animal for tools.
...	...	...	...	...	...	...	...	...	(C).—89, with heads weighing 4½-lbs.; 19, with 8-lb. heads.
1	6	7	...	4	4	1	...	1	(D).—9 pickaxes are carried on each company pack animal for tools.
...	7	7	...	2	2	...	1	1	1 hammer-headed axe is carried on each company pack animal for tools.
1	14	15	...	4	4	1	...	1	
...	4	4	...	2	2	...	3	3	1 hammer-headed axe is carried on one of the spare pack animals.
...	...	...	...	...	...	...	...	...	
...	3	3	...	1	1	...	...	...	(E).—Carried on the person.
...	3	3	...	1	1	...	...	...	
...	54	54	...	18	18	...	...	...	
...	3	3	...	1	1	...	9	9	
...	150	150	...	50	50	...	...	...	
...	3	3	...	1	1	...	...	...	
...	240	240	...	80	80	...	...	...	
...	96	96	...	32	32	...	...	...	

They should be corrected periodically in accordance with the F.S. Manuals various arms.

TABLE OF TIME,

REQUIRED FOR THE EXECUTION

Except where otherwise stated the material and tools are assumed to be the distribution of the working parties at the sites. Not more than five reliefs, if the men have been told off into suitable groups or parties under One leader or foreman can conveniently supervise

No.	Nature of Work.	Reference or Dimensions.	Minutes of One Man.	Per Unit of Task.	Suitable Unit Party.
	ENTRENCHING.				
1	Excavation only	Sec. 18 (3)	3	1 cubic ft	1
	Ditto in small recesses, shelters, etc	—	9	do.	—
2	Fire trench, 1 rifle	Pl. 9, Fig. 1	100	2 paces or 45 cubic ft.	1
3	Fire trench, 1 rifle	Pl. 9, Fig 2	300	2 paces or 90 cubic ft.	1 or 2
4	Fire trench, 1 rifle	Pl. 9, Fig 3	420	2 paces or 110 cubic ft.	1 or 2
5	Communication trench	Pl. 12, Fig. 4	240	2 paces or 80 cubic ft.	1
6	Shovelling loose earth . ..	—	1	1 cubic ft.	1
7	Removing 50 yards (average) deposit, and return ...	by wheelbarrow by stretcher	1 2	1 cubic ft. 1 cubic ft.	1 2
8	Filling sandbags	average ½ cubic ft	3	1 sandbag	3
9	Head cover, sandbags or sods...	Pl. 10, 11	60	1 loophole	1
10	Overhead cover, added to head cover	Pl. 13, Fig. 4	60	1 rifle	1
	REVETMENTS.				
11	Brushwood, rough or planks ...	Pl. 2, Fig. 2	1½	1 square ft.	2
12	Do as hurdle work ..	Pl. 2, Fig. 1	2	1 square ft.	2
13	Sandbag or sack	Pl. 2, Fig. 4	3	1 square ft.	2
14	Gabions, placing and filling ..	Pl. 2, Fig 5.	5	1 square ft	1
15	Sods, building with	Pl. 2, Fig. 3	6	1 square ft.	2
16	Sods, provision of (for above) ...	—	9	1 square ft.	3
17	Gabions, band, making	Pl. 4, Fig. 7	20	1 gabion	2
18	Gabions, brushwood, making	Pl. 4, Fig. 4	360	1 gabion	3

MEN AND TOOLS

of Certain Field Works.

on the site of the work. All tracing and marking out is to be done before minutes should be consumed in distributing the men, or in changing leaders previously instructed in the nature of the particular works in hand. up to 20 unskilled men on earthwork.

Tools per Party.	No.	Remarks and Notes.
1 shovel and 1 pick	1	Averaged over a relief of 4 hours in ordinary easy soil.

Volume excavated in cubic feet. average hourly rates	1st Hour	2nd Hour	3rd Hour	4th to 8th Hours.
Tools used by One Man —	30	25	15	10
Tools Double Manned —	40	33	20	13

Tools per Party.	No.	Remarks and Notes.
" "	2	Including elbow rest and share of drain. If tools are double manned the time can be reduced to 70 minutes.
" "	3	Earth removed or scattered by other men (*see* items Nos. 6 and 7). If tools are double manned and first pair relieved after 2 hours, time can be reduced to 150 mins.
" "	4	This may also be taken as normal for 1 rifle entrenched as Plate 13, Fig. 2, in a recess 3ft. 6in. wide. If tools are double manned and first pair relieved after 2 hours, time can be reduced to 180 mins.
" "	5	If tools are double manned and first pair relieved after 2 hours, time can be reduced to 135 mins
1 shovel	6	As into barrows, boxes, gabions, stretchers, sacks, &c., or cross-lifting or spreading carefully. Averaged over 8 hours work.
1 barrow 1 stretcher	7	2 cubic feet per load. Average weight of earth or sand 1 cwt. per 1 cubic foot. Removing earth over 100 yards is usually more economical by horse and cart, or tram.
2 shovels	8	Averaged over 2 hours work, i e. 120 bags filled by 3 men. Size up to 20in. × 10in. × 5in Weight 60 lbs. For sacks use item No. 6. Corn sacks average 2 bushels = 2½ cubic feet when quite full. Allow 1 cubic foot only.
1 shovel	9	Up to 12 (sandbags or sods), according to description. For spaces between loopholes calculate by items 13 or 15 below, if necessary.
1 shovel, 1 hand axe	10	Allow for 25 square feet of roofing per rifle, in addition to necessary supports. Nails, &c., as obtainable.
1 billhook, 1 mallet	11 12	Allow 4 lbs. of brushwood and 1 foot of wire per 1 square foot of surface revetted.
—	13	Sandbags (already filled) in courses of alternate headers and stretchers 1 sack or 2 bags per 1 sq. ft.
1 shovel, 1 pick	14	Gabions 2ft. wide by 2ft. 9in high; area revetted 5½ sq. ft.; contents 8½ cub. ft. Earth to be excavated.
1 shovel or spade 3 sharp spades	15 16	Allow 5 sods, each about 18in. by 9in. (say 1 sq. ft. each) by 4in. thick, per 1 square foot of *surface revetted* 18 inches thick. Rate of cutting, about 30 sods per hour by one man in 4 hours
—	17	Materials 10 bands and clips, 10 pickets, weight 13 lbs. Filling 25 min. per gabion (*see* No. 14).
1 billhook, 2 knives 1 mallet, 1 measure	18	Materials 75 lbs. brushwood, for use and waste, finished weight about 50 lbs.

TABLE OF TIME.

No.	Nature of Work.	Reference or Dimensions	Minutes of One Man.	Per Unit of Task.	Suitable Unit Party.
	REVETMENTS—*cont.*				
19	Hurdles, rough, making ...	Pl. 5, Fig. 3	60	1 hurdle	3
20	Hurdles, strong, making ..	Pl. 5, Fig. 2	450	1 hurdle	3
21	Fascines, making ...	Pl. 3, Fig. 5	240	1 fascine	4
	CUTTING AND FELLING				
22	Trees, felling	up to 12in. diam	1	1in. of diamr.	1
23	Woods, clearing of brushwood and small trees.	up to 12in. diam.	2½	1 square yd.	20
24	Hedges, felling, stems	up to 2in. diam.	10	1 yard run	2
25	Brick wall, notches in .	up to 18in. thick	10	1 notch	1
26	Brick walls. loopho es in	up to 18in. thick	30	1 loophole	1
	OBSTACLES				
27	Abatis, and wired	1 strong row	120	1 yard run	20
28	Wire entanglement	Pl. 22, Fig. 1, 2	60	1 square yd.	3

MEN AND TOOLS—*continued.*

Tools per Party.	No	Remarks and Notes.
} 2 billhooks, 2 knives } 1 mallet, 1 pr. pliers	19 20	{ Materials 75 lbs. brushwood and 60 ft of wire or yarn per hurdle, 6ft. by 2ft. 9in. Weight of each complete, about 56 lbs
3 billhooks, 2 knives 1 handsaw, 1 maul 1 pr. pliers, 1 choker	21	{ Materials 200 lbs. brushwood and 60 ft of wire or hoop iron (40ft) per fascine, 18ft. long by 9in. diameter. Weight complete about 140 lbs Cradle for making requires 10 pickets, 6ft 6in by 3in diameter.
1 felling axe, or saw	22	{ Over 12 inches diameter allow time in minutes $= \frac{d^2}{144}$, where d = mean diameter in inches. If only hand-axes are available allow twice the time as calculated by both these rules
10 billhooks, 4 felling axes 4 hand axes, 2 saws 1 grindstone, 2 whetstones	23	{ All hands felling at first : then a proportion detailed for collecting and removing according to purpose in view. Produce : about 5 lbs brushwood per 1 square yd For further details *see* M.E., Pt. I., pp 27-30.
1 billhook or hand-axe 1 saw 3 fathoms rope	24	{ Average stiff thorn hedge. If necessary use rope to expose lower stems to the cutting tool.
} 1 pick or crowbar	25 26	{ If possible obtain a mason's chisel and hammer.
as for item 23 : also 2 mauls, 3 pr. pliers 1 pickaxe, 1 shovel	27	{ *See* section 43, p 36. The material must be close at hand. Allow 20 yards wire per yard run per row. The length of the branches is more important than their size. Wire each butt securely to a separate stout picket driven at least 2 feet into the ground. Wire for density near the ground.
1 billhook, 1 handsaw 1 maul, 1 pr. pliers 1 pr. wire cutters 3 rag pads for gripping and straining wire.	28	{ Materials 1 stout post 5 to 9 feet long by 5 to 6in. diam per 4 square yards of obstacle. 1 stout picket 2 to 3 feet long by 5in diam. per yard run of finished obstacle, for side stays 10 to 15 yards of wire per 1 square yard of finished obstacle.
In hard ground add : 1 steel jumper 1 sledge hammer		N.B.—A very formidable obstacle can be made with 3 stout posts, 1 stout picket and 150 yards wire per yard run complete. Barbed wire is most quickly handled and fixed if issued in lengths of about 10 feet, when used in entanglements. This does not apply to straight fencing such as shown in Pl. 22, Fig. 8.

APPENDIX III.

Principal Tools, Materials, and Stores suitable for use in Field Engineering.

The tools and stores provided for the Peace instruction of troops in Field Engineering are as laid down in the Regulations for the Equipment of the Army, Part I, 1909, para. 328 and Appendix VI. The tools and stores forming war equipment of units are similarly detailed in the various sections of Part II, Equipment Regulations, and in Mobilization Store Tables.

The following tables of Tools, Materials, and Stores are intended as a guide for the selection and preparation of articles, suitable for use in War, for such operations of Field Engineering as are described or indicated in this Manual.

The method of obtaining supply of such articles will follow the instructions laid down in F.S. Regulations, Part II, Secs. 36 and 52, and Ordnance Manual (War) paras 86 and 87.

The special equipment required for the following Engineer Services is not included in these tables, except in so far as certain articles comprised therein may be suitable for general Field Engineering purposes :—

> *Balloon and Kite Stores.*
> *Electrical instruments and Electric Light Stores.*
> *Railway tools, plant, and Armoured Trains.*
> *Survey instruments and stores.*
> *Telegraph and Telephone Stores.*

Articles in these tables marked N.I.V. are not in the Priced Vocabulary of Stores sealed for use in His Majesty's Land Service; and supply, if required, must be obtained as above directed.

Articles marked N. are a Naval Store Supply.

Demand and Issue, except where otherwise stated, are ' per article.'

TABLE 1.—TOOLS, ENTRENCHING.

DESIGNATION.	DETAIL.	ISSUE.
Axes, pick, heads...	4½ lb. and 8 lb.	Each.
,, ,, helves	36 in. ferruled	,,
Barrows, hand, double ...	6 ft. 7 in. long	,,
,, wheel, entrenching ...	steel tubular frame	,,
Crowbars, chisel and claw ends	6 ft. 37 lb., 5 ft. 6 in. 31 lb., 4 ft. 6 in. 20 lb., 3ft. 6 in., 12 lbs. 2 ft. 3 in. 7 lb.	,,
,, ,, ,,		,,
Implements, entrenching, 1908.	steel, shovel and pick head ...	,,
,, ,, helves	ferruled end	,,
Picks, miners	22¼ in. 6 lb. (special short for cramped work)... ...	,,
,, push	30 in. 3 lb. 6 oz. (heart shaped, straight stabbing) ...	,,
Shovels, G.S.	32 in. helve, 3½ lb.	,,
,, R.E.	32 in. helve, 5lb.	,,
,, miners	30 in. 6 lb., and long 5 ft. ...	,,
Spades, Mark III	32¾ in. helve, 5¾ lb.	,,
Plates, loopholed, Mark II ...	steel, 2 ft. × 1 ft. × ⅟₁₆ in... ...	,,
Shields, sap	steel, 1 ft. 11 in. × 1 ft. 9 in. × ⅟₁₆ in. 25¼ lb.	,,

TABLE 2.—TOOLS, CUTTING

Designation.	Detail.	Issue.
Adzes, carpenters, handled ...	4½ lb.	
Axes, felling	32 in. helve, 6 lb. 7 oz.	
Axes, hammer headed	pioneers, 16 in. helve.	
Axes, hand	16 in. helve, 2 lb. 3 oz.	
Chisels, brick	18 in., 1¼ in. end.	
Chisels, firmer	wood-cutting, blades 3 in. to $\frac{1}{16}$ in. wide.	
Chisels, hand, cold	metal cutting, 1 in., ⅞ in. and ¾ in. wide.	
Cutters, expanded metal ...	size 3, powerful, to cut steel rod and rope to ⅜ in. diam.	prs. N.I.V.
Grindstones, F.S. .. .	18 in., 76 lb. 10in., 25 lb.	
Hooks, bill	1 lb. 13 oz.	
Hooks, reaping	1 lb.	
Pliers, side-cutting ...	8 in. and 5 in. long	prs.
Rods, clearing obstacles ..	bamboo, in four 5 ft. lengths, with saw, shears, hook and line.	
Saws, cross-cut	5 ft. blade, 6¾ lb.	
Saws, folding, in leather case ...	3ft. 9in. blade, 2 handles. 1lb. 12oz.	
Saws, hand	26 in. and 20 in.	
Sets, cold, large	15 in. handle; for cutting steel wire rope, &c.	
Stones, rag ...	for reaping hooks, &c.	doz
Wirecutters, in frogs, Mark IV.	9½ in. long, 1¼ lb.	

TABLE 3—TOOLS, ARTIFICERS, COMPLETE SETS.

Designation.	Detail.	Issue.
Chests, Tool, Filled :—	For contents *see* Equip. Regns., Pt. I, App. V, p. 145.	
Bricklayers and masons ...	No. 2 chest	each
Carpenters and wheelers ...	No. 3 chest.	
Carpenters	No. 4 chest.	
Coopers	No. 6 chest.	
Farriers and shoeing-smiths	No. 7 chest.	
Painters and glaziers ..	No. 13 chest.	
Plumbers and tinsmiths .	No. 12 chest.	
Smiths	No. 11 chest.	
Tools, screw-cutting, bolt and nut, Whitworth standard thread, Mark III	Set A, 1½ to 1⅛ in. Set B, 1 to ⅝ in. Set C, ½ to $\frac{5}{16}$ in. Set. D, ¼ to ⅛ in.	sets „
Tools, screw-cutting, iron and steel tube, Mark II	Set, 1⅛ in. to ⅛ in.	„

TABLE 4.—TOOLS AND STORES, MISCELLANEOUS.
[Not Included in Tables 1, 2 or 3.]

DESIGNATION.	DETAIL. [Sizes selected for general Field use.]	ISSUE.
Anvils	1 cwt.	Each.
Barrows, drum, universal ...	for paying out wire.	
Bars, boring (steel chisels) ...	for rock, 3 in., 1½ in., and 1¼ in. wide, up to 4 ft. long.	
Bars, jumping (chisel each end)	3 in., 1½ in., and 1¼ in. wide, 5 ft. 6 in. and 7 ft. long.	
Bars, pinching (spike and lever)	2½, 3, 3½ and 4 ft. long.	
Bells, bicycle	suitable for alarm signals, &c.	
Blocks, tackle, G.S. cast iron galv.	single, double, treble, and snatch (and size of cordage).	
Blowers, rotary, Mk. IV	with hose and wrenches; for ventilating mines, &c.	
Buckets, miners	14lb.7oz.for raising earth from shafts	
Candlesticks, miners	with bottom and side spikes.	
Carts, trench, Mark II	for man draught.	
Chevaux de frise, iron	1 barrel (6 ft.) and 12 spears (5 ft. 5½ in.)	
Chokers, fascine	2 four ft. levers and 4 ft. chain.	
Climbers, pole or tree	steel, with 4 straps and 2 pads ...	páir.
Clips, lifting, iron	with 5 ft. ash shaft; for lifting steel rails, &c.	
Crabs, hoisting, iron	hand power winches, to lift 1, 25, and 50 tons.	
Foghorns, mechanical, brass ...	large: small. For alarms, signals, &c.	
Forges, field, G.S. ...	276 lb. and poker, slice, tongs and vice	
Forks, sandbag	4 ft. long; 5 lb. 10 oz.	
Grapnels, iron	2, 3, 16, 40, and 50 lb.	
Hammers, claw	20 oz.	
Hammers, masons	10 lb. chisel point.	
Hammers, miners, boring ...	5 and 7 lb.	
Hammers, miners, sledge ...	14 lb.	
Hammers, R.E. Telegraph, sledge	16 lb.; for driving steel jumper.	
Jumpers, steel	for post holes, 2 ft. 9 in. × 2 in. 27¾ lbs.	
Ladders, field telegraph	16 ft. 6 in. in 2 lengths.	
Ladders, rope, miners	20, 30, and 50 ft.	
Ladders, scaling	in lengths; to join; 8½ ft. and 6 ft. each	
Levels, F.S.	4 ft. 3 lb. 7 oz.	
Mallets, heel peg, Mk. III ...	head 6¾ in. × 4 in.	
Mauls, G.S.	14 lb.	
Megaphones, papier mâche ...	large; small.	
Rods, measuring, common ...	wood, 10 ft. and 6 ft. marked 3 in. 5 ft. marked ⅛ in.	
Scoops and scrapers	for clearing bore holes; 3 in., 1½ in., and 1¼ in. × 6 ft. 6 in.	
Scrapers, earth	7 ft. long; 3 lb.; also with 3 ft. handle.	
Spanners, adjustable	15 in.	
Spanners, McMahon ...	9 in.	
Spurs, pole, ¾ in.	for spiking to trees, &c. to form steps.	
Tapes, measuring	in leather case, 100 ft.	
Tapes, tracing	50 yds.; 1½ in. white web.	
Trucks, miners, elm	69 lbs.; for removing earth in saps and mines.	
Vices. standing. 36 lb.	jaws 4 in. wide.	

TABLE 5.—BRIDGING AND BOAT STORES.

[*See also* MILITARY ENGINEERING, PT. III.]

DESIGNATION.	DETAIL.	ISSUE.
Anchors, boat	1 cwt., and ½ cwt.	
Bailers, pontoon	tin, with handle.	
Barrels and casks...	*See* Equip. Regns., Pt. I, p. 174; and Sizes, Sec. 97 (4) above ...	N.I.V.
Baulks, barrel pier, Mk. III ...	15 ft. × 4 in. × 4 in.; 55 lb.	
Baulks, Mk. III, tapered ...	15 ft. 9¾ in. × 3¼ in. to 1½ in × 6 in. 56 lb.	
Baulks, shore end, inside ..	3 ft. 7 in. long; 3 to set; 15½ lb. each.	
Baulks, shore end, outside ...	3 ft. 6¼ in. long; 2 to set; 21½ lb. each.	
Beams, saddle, Mk. II	in two pieces; 58 lb. pair.	
Boats, collapsible ...	bow and stern sections; 6 ft. 1½ in. long; 9 ft. oars.	
Buoys, pontoon, iron ...	for anchors; 5 lb.	
Chalk, prepared	white or coloured; 144 pieces ...	Box.
Chesses, Mk. II	10 ft. × 12 in. × 1½ in.; 45 lb.	
Gunnels, barrel pier	21 ft. × 5 in. × 4 in.; 115 lb.	
Drivers, pile, Swiss	with iron guide rod, about 130 lb., hand power.	
Hooks, boat	18 ft., 11 ft. 7⅜ in., and 6 ft. long.	
Life-belts, cork.		
Life-buoys, Mk. IV	reindeer hair, covered canvas.	
Oars, ash	20 ft. to 8 ft. long, by 1 ft. difference. 12 ft. for pontoons.	
Outriggers, barrel pier, Mk. II	11 ft. 6 in. × 4½ in. × 3 in. 25 lb.	
Pontoons, bipartite, Mk. II .	bow and stern pieces. 1,008 lb. per pair.	
Ribands, Mk. II	15 ft. 9 in. × 3¼ in. × 6 in. 79 lb.; can be used as baulks.	
Stay tighteners, screw, ⅝ in. G. I	straining screws for wire slings, stays, &c.	
Sticks, rack	with 6 ft. of 2 in. lashing. 1¾ lb.	
Transoms, shore end, Mk. III ...	11 ft. 6 in. long. 73 lb.	
Travellers, rope, steel ...	2 pulley wheels in frame, for flying bridges, ferries, &c.	N.I.V.
Trestles, bridging, Mk. III ...	with 2 tackles, differential, 10 cwt. 816 lb.	

For lashings and wire rope, *see* Tables 6 and 14.

TABLE 6.—CORDAGE.

DESIGNATION.	DETAIL.	ISSUE.
(i) Cordage, hemp, hawser, 3-strand	Service cordage in general use; either *tarred* or *white*; in the following sizes, circumference in inches :—9, 7, 6, 5, 4, 3½, 3, 2½, 2, 1½, 1 in. *Tarred* cordage is weaker, but will stand exposure to weather better than *white*. A special size, 1¾ in., is issued in lengths of 66 ft. for picketing ropes.	fathom [coils of 113 fms.]
(ii) Cordage, manilla, hawser, 3-strand	A stronger cordage, suitable for drag-ropes, head and heel ropes, &c., in the following sizes :—5, 4, 3½, 3, 2½, 2, 1½, 1, ¾, ½ in.	"
(iii) Cordage, coir, hawser, 3-strand	A coarse, light, elastic cordage, which will float upon water, but has only one-sixth the strength of hemp cordage of same size. Sizes :—9, 7, 6, 5, 4, and 2½ in.	"
(iv) *Lashings, falls, guys, &c.* ...	Cordage, as in (i) above, of sizes as under :— 3 in. Footropes, 9 fms.; cables, 30 fms.; falls, 50 fms.; guys, 30 to 36 fms. 2½ in. Slings for cask piers, 6 fms. 2 in. Falls, 50 fms.; lashings, 6 and 9 fms. 1½ in. Braces, 3 fms.; breast lines, 10 fms.; lashings, 6 fms. 1 in. Buoy lines, 10 fms.; lashings, 3 and 6 fms.	"
(v) *Small cordage, yarn, twine, &c.*— Cordage, spun yarn, hemp ...	3-thread, tarred, rough	cwt.
Cord, sash, ½-in.	plaited, as for windows, &c. ...	doz. yds.
Lines, bricklayers' and masons'	20 yds., strong and light	doz.
Lines, Hambro	150 ft. " "	each.
Line, log, patent plaited ...	No. 9 " "	fm.
Line, tarpon, 24-thread, stout	very strong and light; for trip lines, flares, alarm signals, &c. 200 yards on reel	N I.V.
Lines, sounding, 1-in.	25 and 42 fms.	each.
Line, tarred	⅛ and ¼ in.	lb.
Line, tracing	for tent lashings and tracing lines	lb.

TABLE 7.—DEMOLITION STORES AND EXPLOSIVES.

DESIGNATION.	DETAIL.	ISSUE.
Blasting gelatine	plastic explosive, 50 % stronger than dynamite	N.I.V.
Boxes, testing and jointing, filled	for preparing to fire charges electrically. 12 lb. 13 oz. ...	each.
Cables, electric, E. 1, Mark II	for field demolitions. 57 lb. per 1,000 yards ...	yards.
Cables, electric, D. 5, Mark IV	for field demolitions, lighter. 48lb. per 1,000 yards	yards.
Caps, copper, blasting	detonators for dynamite, &c., commercial, sizes 3 to 10	100 N.I.V.
Circuit closers, pull or thread ...	for firing land mines, &c., M.E., Pt. IV, Plate 21 ..	N.I.V.
Cordite, size 3	as for small arms, on drums ...	lb.
Detonators, No. 8, Mark V ...	for instantaneous and safety fuze. Tins of 25 ...	100.
Do. with 2 ft. of safety fuze attached	for cavalry pioneers, &c. Tins of 6 with rectifier	100.
Detonators, electric, No. 13, Mark III	for firing high explosives electrically	100
Dynamite, No. 1	commercial explosive; not waterproof	N.I.V.
Exploders, dynamo electric, Mark V	for firing charges electrically. $13\frac{1}{2} \times 8\frac{3}{8} \times 6\frac{1}{4}$ in. 27 lb. ...	each.
Fuzes, electric, No. 14, Mark III	for firing gunpowder electrically. Tins of 25	each.
Fuze, instantaneous, Mark III	in zinc boxes of 100 yds.	100 yds.
Fuze, safety, No. 9, Mark II ...	in tin cylinders of 8, 24 and 50 fms.	100 fms.
Grenades, hand, percussion ...	brass head and cane handle. 1 ft. $9\frac{3}{4}$ in. 1 lb. 12 oz.	each.
Grenades, rifle, percussion ...	to fire from rifle	N.I.V.
Guncotton, dry, primers, field, 1 oz.	conical, 1·35 to 1·15 in. diam. × 1·25 in., 1 perforation	each.
Guncotton, wet, charges, field, 15 oz.	one slab in copper-tinned case, 1 perforation for primer, 1 oz....	each.
Guncotton, wet, slabs, field, 15 oz.	6 × 3 × $1\frac{3}{8}$ in.; 1 perforation for primer, 1 oz.	each.
Gunpowder, L.G. or R.L.G. ...	per barrel of 100 lb.; or $\frac{1}{2}$, $\frac{1}{4}$ and $\frac{1}{8}$ barrel	100 lb.
Match, quick	prepared cotton wick; burns 1 yd. in 13 secs.	lb.
Match, slow	prepared hemp; burns 1 yd. in 8 hours	lb.
Matches, Vesuvian	fuzees; 20 in box	doz. boxes
Pistol, B.L. safety fuze, Mark IV	and cartridges; will ignite safety or instantaneous fuze	N.
Rectifiers, guncotton, primers, Mark V	for enlarging detonator hole in primers, field, 1 oz.	each.
Tubes, friction, copper, solid drawn, Mark I.	for firing gunpowder or detonator No. 8	100.

TABLE 8.—LIGHTS, LAMPS, FLARES, ROCKETS, Etc.
(Exclusive of Army Signal Apparatus.)

Designation.	Detail.	Issue.
Lamps, acetylene	hand and portable; and calcium carbide	N.I.V.
Lamps, bicycle	oil ..	each.
Lamps, electric	hand and portable; and spare accumulators	N.I.V.
Lamps, hurricane ...	oil or candle	each.
Lamps, miners	oil; 3 glass sides	each.
Lamps, railway guards ..	oil; clear, blue, and red lights ...	each.
Lamps, tracing	and signalling; bull's-eye with flashing disc	each.
Lamps, Wells' flare, petroleum .	No. 1, approx. 500 candle power; 1 gall. per hour. 70 lb. filled ...	N.I.V
Do. do. do. ..	No. 3, approx. 900 candle power, 1¾ gall. per hour. 280 lb. filled	N.I.V.
Lanterns, coloured	square, 3 bull's-eyes, siege batteries and R.E.	each.
Lanterns, field telegraph, Mk. II	oil or candle; tin and talc sides ..	each.
Lights, illuminating wrecks ...	with stand; composition burns about 25 min.	N.
Lights, coastguard, Mk. II ...	each with spike; composition burns about 5 min.	N.
Lights, long, blue, Mk. III ...	wood handle; also green and red, 5 min.	each.
Lights, long, G.S., Mk. III ...	wood handle, plain, 5 min. ...	each.
Pistol, signal, Very cartridge ...	and cartridges; fires green, red or white 9 sec. star 300 ft.	pistol, each; cartridges 100.
Portfires, common	16 in. long. In bundles of 12; burn 12 min.	each.
Portfires, life saving	8 in. long. Tins of 25	each
Rockets, flash and sound, 1 lb. ..	and 5 ft. stick. To insert: "Detonator, sound rocket, No. 2," and "Guncotton, primer, rocket, 2 oz."	each.
Rockets, light and sound, 1 lb....	and 5 ft. stick, with notch. To insert: "Charge, tonite," and "Detonator, S.R., No. 2" ...	each.
Rockets, light, ½ lb.	and 4 ft. 2 in. stick. Single 15 sec. magnesium star	each.
Rockets, signal, 1 lb., service ..	and 5 ft. stick, with notch; also blue, green or red; 28 stars ...	each.
Rockets, signal, 1 lb., 'red and white'	and 5 ft. stick, with notch; 25 red and 24 white stars	each.
Rockets, signal, ½ lb., service ...	and 4 ft. 2 in. stick; also blue, green or red; 20 stars	each.
Rockets, sound, ½ lb., Mk. III..	and 4 ft. 6 in. stick. To insert . "Charge, tonite," and "Detonator, S.R., No. 2"	each.
Charges, tonite	say whether for "1 lb. light and sound," or "½ lb. sound" ...	each.
Detonators, sound rocket, No. 2	for sound, or light and sound rockets	100.
Guncotton, primers, rocket, 2 oz.	for flash and sound rocket, 1 lb. ...	each.
Signals fog	as alarm signal; to be crushed or struck	dozen.

TABLE 9.—BOLTS, DOGS, NAILS AND SPIKES.

DESIGNATION.	DETAIL.	ISSUE.
Bolts, with nuts, hexagon head ...	principal store sizes; length and diam. in inches 14 × 1 or ¾. 12 × ¾ or ⅝. 8 × ¾ or ⅝. 6 × ⅝ or ½. 5 × ⅜. Other sizes prepared as required.	100
Bolts, eye, straining, galv. iron	for straining wire; or light timber fastenings; trade sizes:—18 × ⅝. 15 × ⅝ or ½. 12 × ½ or ⅜. 9 × ⅜ in.	N.I.V.
Dogs, railway and sawyers Mk. II ...	straight. 15 and 12 in. long, with 6 in. teeth ...	each.
Nails, iron spike (*quote Store No.*)	Length inches:— 10 | 9 | 8 | 7 | 6 | 5 Nails in 1 cwt. (app.):—114 | 155 | 193 | 294 | 430 | 590 Army Store No.:— *187* | *186* | *185* | *184* | *183* | *182*	cwt.
Nails, wire, iron, grooved	Length, inches:—6 | 5 | 4 | 3 | 2½ | 2 | 1¾ | 1½ | 1¼ | 1 Nails in 1 lb. (approx.):— 14 | 20 | 50 | 70 | 100 | 150 | 200 | 300 | 400 | 600	cwt.

TABLE 10.—POSTS, PICKETS, ETC.

DESIGNATION. DETAIL:—	LENGTH. ft.	LENGTH. in.	DIAMETER. inches.	ISSUE.
Pegs, stay, Mk. II, steel	1	1	— 2½ lb.	each.
Pickets, brushwood, high wire entanglement	5	0	3 to 4	"
Pickets, brushwood, gabion	3	6	¾ to 1	100
Pickets, brushwood, low wire entanglement	2	6	1½ to 2	each.
Pickets, brushwood, tracing	1	6	1 to 1¼	100
Pickets, gabions, band	3	0	1¼ × ½	100
Pickets, square, high wire entanglement	5	0	3 × 3	100
Pickets, square, low wire entanglement	2	6	2 × 2	100
Pickets, fascine cradles	6	6	3 to 4	N.I.V.
Pickets, hurdles	3	6	1 to 2	N.I.V.
Posts, wire entanglement, steel... ...	5	6	with 6 screw hooks and nuts	each.
Posts, wire entanglement, wood, up to	8	0	5 to 6	N.I.V.
Posts, picket, 8 ft. Mk. IV	8	0	5⅞ to 5¼ tapering ...	each.
Posts, picket, 5 ft. Mk. III	5	0	3 to 2⅜ tapering ...	"
Posts, picket, 3½ ft. Mk. II	3	6	2½	"
Posts, picket, 2½ ft., Mk. IV	2	6	2⅜ to 2 with steel staple	"
Posts, picket, 2½ ft., Mk. V	2	6	2⅜ to 2 with rope loop	"

TABLE 11.—SANDBAGS, SACKS, CANVAS, COVERINGS, TARPAULINS, ETC.

DESIGNATION	DETAIL.	ISSUE.
Bags, ammunition, canvas, Mk. V	5 ft. 10 in. long	each.
Bags, corn, 8 lb.	canvas, 2 ft. 4½ in. × 9 in. ...	"
Bags, corn, 2 bushels	canvas, 4 ft. 3 in. × 1 ft. 4½ in. ...	"
Bags, gun cotton, waterproof ..	canvas, to hold 2, 5, and 25 lb. ..	"
Bags, pin, tent, canvas ...	1 ft. 10 in. × 1 ft. 5 in., 1 ft. 11 in. × 1 ft. 10 in., 2 ft. 6 in. × 2 ft. 6in.	"
Bags, powder barrel, waterproof	to hold 100, 50, 25, and 12½ lb. ...	"
Bags, sand, common, Mk. II ...	canvas, 2 ft. 9 in. × 1 ft. 2 in. Bales of 100 ; 33 to 43 lb.	"
Canvas, bleached, waterproof ...	40 inches wide	yards.
Canvas, rot proofed	60 inches wide	"
Canvas, tanned	24 inches wide (stretchers, &c.) ...	"
Canvas, tarred	40 inches wide	"
Canvas	36 inches wide (wagon covers, &c.)...	"
Covers, sail cloth, waterproofed	sizes in feet:—30 × 30 ; 30 × 20 ; 24 × 18 ; 20 × 16 ; 18 × 15 ; 15 × 10 ; 12 × 10	each.
Sacks, coal, 2 cwt.	tarred or untarred, 4 ft. 5 in. × 2 ft. 2 in., with beckets	"
Sacks, coal, 100 lbs., tarred ...	3 ft. × 2 ft., with beckets	"
Sacks, coal, packsaddle, ½ cwt. ...	3 ft. × 1 ft. 4 in.	"
Sacks, corn, canvas	4 bushels	"
Sacks, jute	5 ft. × 3 ft., 4 ft. 6 in. × 2 ft. 4 in. ...	"
Sheeting, corrugated galv. iron	26 in. wide, 5 to 12 ft. long ; 24 gauge ; 840 ft. per ton ...	N.I.V.
Sheets, ground, waterproof ...	6 ft. 6 in. × 3 ft.	each.
Tarpaulins	sizes in feet:—30 × 30 ; 30 × 20 ; 30 × 16 ; 24 × 18 ; 20 × 16 ; 20 × 10 ; 18 × 15 ; 15 × 10 ; 12 × 10 ; 10 × 6	"

TABLE 12.—TIMBER.

Note.—For hasty field engineering in war it is only possible, as a rule, to make use of such timber as may be available close to the site of the work.

Such timber will either be felled and trimmed on the spot, collected from timber stores in adjacent towns and villages, or obtained by dismantling structures containing timber, such as sheds, floors, gates, doors, vehicles, etc.

In countries where timber is very scarce, too hard or too heavy, or otherwise unsuitable, and when time permits of obtaining supply of timber from a distance, the following notes on the natures and limiting sizes of timber in ordinary commercial use may assist in preparing demands for material suitable for general field purposes in war.

DESIGNATION.	DETAIL.	ISSUE.
BAULK TIMBER—	Logs trimmed and roughly squared.	
American yellow pine ..	up to 70 ft. × 24 in. diam. ...	ft. cube of
Baltic fir, or equal	up to 25 ft. × 12 in. „ ...	given size
SAWN SCANTLINGS—	'Baltic deals,' or equal, usual market sizes:—	
Battens	up to 22 ft. × 4 to 7 in. wide × 1 to 4 in. thick.	in sizes as demanded;
Deals	up to 22 ft. × 8, 9, 10 in. wide × 1 to 4 in. thick.	length in ft., width
Planks	up to 22 ft. × 11, 12 in. wide × 1 to 4 in. thick.	and
	The most generally useful and easily obtained sawn scantlings, are 'Baltic deals,' 9 × 3 in., in lengths from 8 to 20 feet, but chiefly 12 ft.	thickness in inches.
Skids, pitch pine	as stored for ordnance services; suitable for heavy timber road and railway bridges, &c.	
	sizes:—30 ft. × 18 × 18 in.; 20 ft. × 15 × 15 in. 14 ft. × 12 × 12 in.; 6 ft. × 12 × 12 in.	each.
Timber, flooring	10 ft. × 9 in. × 1¼ in., for bridging purposes. [poses	each.
„ planks, bridging service	10 ft. × 12 in. × 1¼ in. .., „ pur-	„
„ poles, hutting ...	in lengths and sizes as required ...	—
„ sheeting, miners ..	5 ft. × 11 in. × 2 in.; fir	piece.
„ spars, bridging service	in lengths and sizes as required ..	—
„ spars, sheers and derricks	sizes:—for sheer legs:—70 ft. × 31 in., 60 ft. × 20 in., 50 ft. × 15 in., 40 ft. × 13 in., 40 ft. × 10 in., 30 ft. × 10 in.	each.
	for derricks:—30 ft. × 9 in., 25 ft. × 9 in., 25 ft. × 7 in. and 20 ft × 9 in.	each.

TABLE 13.—WATER SUPPLY STORES.

DESIGNATION.	DETAIL.	ISSUE.
Buckets, water, G.S., Mk. III .	canvas, 3 gallons	each.
Buckets, well, tip, iron	27 gallons	"
Cans, water	3 gallons	"
Engines, oil, 8 B.H.P.	for driving "pumps, horizontal, 3,600 gall."	"
Filters, Brownlow, F.S. ...	filters, 25 gall. per hour	N.I.V.
Flags, distinguishing, water supply.	blue, red, and white; and poles 10 ft. in 2 parts.	each.
Hose, canvas	3 and 4 in. diam.	yards.
Hose, delivery, canvas, 2¾ in. ...	in lengths 100, 50, 30, 20 ft., with screw unions.	lengths.
Hose, delivery, 2¾ in.	in 30 ft. lengths; prepared for pump L. & F., Mk. IV.	"
Hose, suction, 2 in.	in 12 ft. lengths; prepared for pump L. & F., Mk. IV.	"
Hose, suction, syphon, with cap	3 in. and 2½ in. in 10 ft. 3 in. lengths; 2 in. in 10 ft. lengths.	"
Pails, iron galvanized	3 and 4 gallons	each.
Pumps, deep well	50 ft., 3 in. bore; and 100 ft., 3 and 4 in. bore.	sets.
Pumps, horizontal, 3,600 gall., 4 in.	to raise 3,600 gall. per hour 300 ft. with 'engine, oil, 8 B.H.P.'	each.
Pumps, lift and force, Mk. IV	with four 12 ft. lengths of suction, and 30 ft. of delivery hose, to lift 60 ft. ; weight 84 lb.; with hose 216 lb.	"
Pumps, steam, portable, Merry-weather.	small 'Valiant' on wheels, 8½ cwt.; to raise 1,500 gall. per hour 250 ft.	"
Tanks, steel, corrugated, gal-vanized, circular.	25, 50, 100 to 1,000 gallons; with taps and covers.	N.I.V.
Tanks, waterproof, 2,300 gall., open.	16 ft. 9 in. × 16 ft. 9 in., with stores as M.E., Pt. V, para. 92.	each.
Tanks, waterproof, 1,500 gall., closed.	octagonal; no extra stores re-quired.	"
Tools, boring, artesian ...	see M.E., Pt. V., para. 139, and for tube wells, para. 123.	—
Troughs, waterproof, 600 gall.	with standards, 10 to a set, to water 16 animals at one time.	sets.
Windlasses, well, light	with large drum	each.

Note:—For the transport of water by cart, pack animal, &c., see Sectio.s 2 and 21 Vocab. of Stores.

TABLE 14.—WIRE AND WIRE ROPE.

DESIGNATION.	DETAIL.	ISSUE.
Rope, galvanized, steel, wire .	in coils of 100 fms.	fms.

Sizes, circumference, inches ...	12	11	10	9	8	7	6	5	4	3	$2\frac{1}{2}$	2	$1\frac{1}{2}$	$1\frac{1}{4}$	$1\frac{1}{8}$	1	Sizes 12 to 5 in. are N.I.V.
Approx. weight, lbs. per fm.	115	97	80	65	53	41	33	$23\frac{1}{2}$	12	7	$4\frac{1}{2}$	$2\frac{3}{4}$	$1\frac{3}{4}$	1	$\frac{7}{8}$	$\frac{3}{4}$	
Safe load ($9c^2$) cwt.	1296	1089	900	729	576	441	324	225	144	81	56	36	20	14	11	9	

DESIGNATION.	DETAIL.	ISSUE.
Rope, steel, ·65 in.	for use with collapsible boats, 2 to $2\frac{1}{2}$ tons breaking strain; ·42 lb. fm.	100 yds.
Rope, wire, towing, steel, ·67 in.	1,760 yards on iron reel	per length
Staples, No. 8, S.W.G.	galvanized; about 51 per lb. for fencing, &c.	lb.
Wire for use in wire obstacles:—		
Wire, galv. iron, No. 14 S.W.G.	plain, 100 lbs. per mile; length in 1 cwt., 1,970 yds.	cwt.
Wire, galv. steel, No. 14 S.W.G., 2-strand, barbed ...	280 lbs. per mile; length in 1 cwt., 700 yds.	cwt.
Wire, galv. steel, No. 18 S.W.G., 2-strand, barbed ...	135 lbs. per mile; length in 1 cwt. 1,460 yds.	cwt.
Wires suitable for fencing:—		
Rope, galv. steel wire ..	$\frac{3}{4}$, $\frac{1}{2}$, and $\frac{3}{8}$ in. circumference ..	fm.
7-ply galv. strand wire ...	No. 1, 205 yds.; No. 2, 232 yds.; No. 3, 260 yds. per cwt. Cannot be cut by ordinary hand wire cutters	N.I.V.
Heavier barbed wires:—	commercial sizes.	
'Thickset 4-point galv. steel barb fencing wire, barbs 3 in. apart'	440 lb. per mile; length in 1 cwt. 448 yds.	N.I.V.
'Ordinary 2-point galv. steel barb fencing wire, barbs 5 in. apart'	335 lb. per mile; length in 1 cwt. 589 yds.	N.I.V.
Wires in common use for telegraph air lines:—	Bronze 40 and 70, and (3-strand) 100 lbs. per mile ... ... Copper 100, and galv. iron 200 and 400 lb. per mile	lb.
Wire, ordinary, steel, or iron	(*if steel say 'hard' or 'soft'*) is issued in sizes from 1 to 26 of the standard wire gauge ...	cwt.

TABLE 14.—WIRE AND WIRE ROPE—*continued*.

The table below gives the properties, weight, &c., of new iron wire.   New steel wire may be taken as twice the strength given, otherwise similar in size, &c., galvanized wire is heavier.

Size, S W G.	1	2	3	4	5	6	7	8	9	10	11	12	13
Diam., inches ...	·300	·276	·252	232	212	·192	·176	·160	·144	128	116	104	092
Yards, per cwt ...	155	183	220	260	311	380	452	546	675	854	1040	1298	1658
Lbs per mile ...	1268	1073	895	758	633	518	436	360	292	231	190	152	119
Approx. breaking strain, lbs.	3804	3219	2685	2274	1899	1554	1308	1080	876	693	570	456	357

Size, S.W.G.	14	15	16	17	18	19	20	21	22	23	24	25	26
Diam., inches ...	·080	·072	·064	·056	048	·040	·036	·032	028	·024	·022	·020	018
Yards, per cwt ...	2186	2699	3416	4462	6078	8745	10796	13668	17846	24290	28908	34978	43184
Lbs. per mile ...	90	73	58	44	82·5	22·5	18·2	14·4	11	8·1	6 8	5·6	4·6
Approx breaking strain, lbs.	270	219	174	132	98	68	55	43	33	24	20	17	14

INDEX.

Plate 1

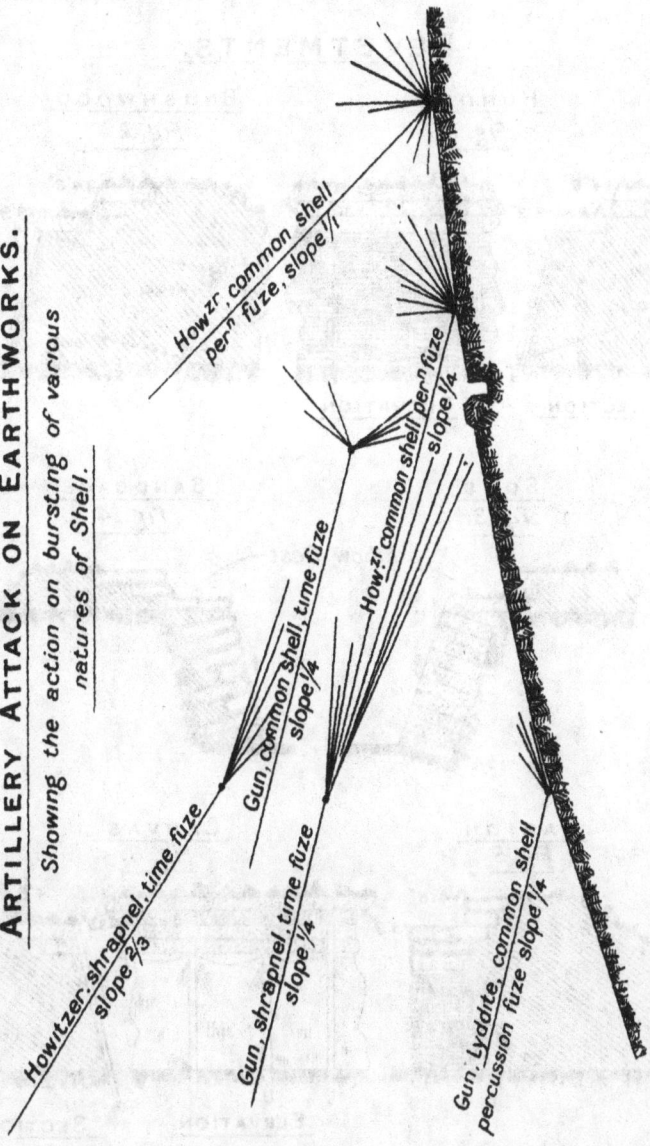

ARTILLERY ATTACK ON EARTHWORKS.

Showing the action on bursting of various natures of Shell.

Howitzer, shrapnel time fuze slope 2/3

Gun, common shell, time fuze slope 1/4

Gun, shrapnel time fuze slope 1/4

Howr, common shell perᶜ fuze, slope 1/1

Howr. common shell perⁿ fuze slope 1/4

Gun, Lyddite, common shell percussion fuze slope 1/4

Plate 2

REVETMENTS.

HURDLE
Fig. 1.

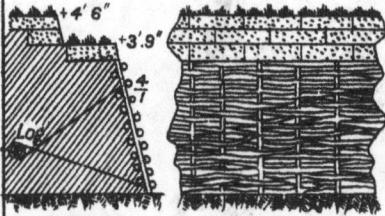

+4'6"
+3'9"

Log

SECTION   ELEVATION

BRUSHWOOD
Fig. 2

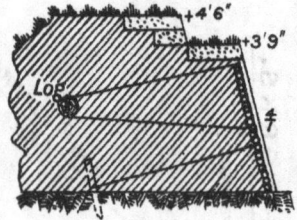

+4'6"
+3'9"

Log

SODS.
Fig. 3

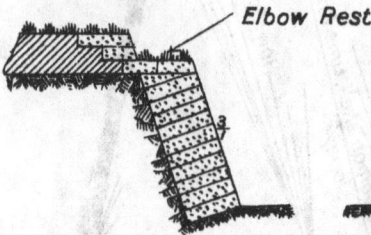

Elbow Rest

SANDBAGS
Fig. 4.

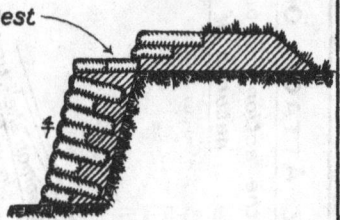

GABION
Fig 5

+4.6"
+3'9"

CANVAS
Fig. 6.

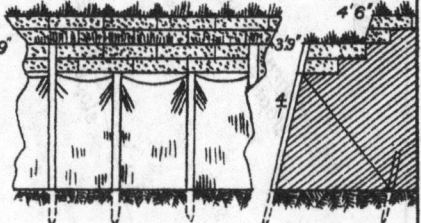

4'6"
3'3"

ELEVATION.   SECTION

471 8608/183 90000 7 11   Malby & Sons, Lith

Plate 3

FASCINES.

WITHES.

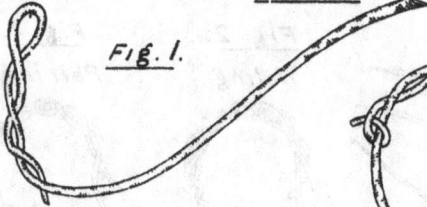
Fig. 1.

Fig. 2.

Fig. 3.

FASCINE TRESTLE.

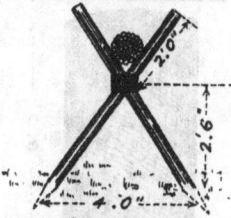

*Stakes
about 6½' long*

Fig. 4.

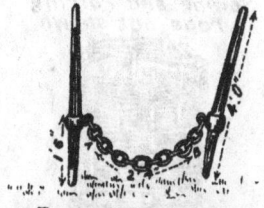

FASCINE CHOKER.
For compressing brushwood.

18 F.T FASCINE   *Trestles 4' apart*

Fig. 5.

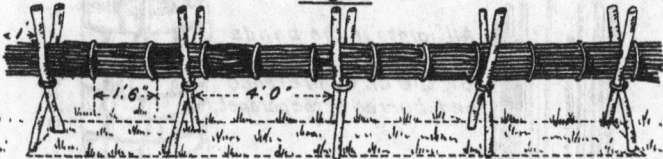

Suitable tracing rectangle measures 16' 0" × 4' 0"

471 9609/185 30000.7.11.   Malby & Sons, Lith

Plate 4

GABIONS.

Fig 1
Commencement of Waling

Fig. 2.
Waling

Fig. 3.
Pairing

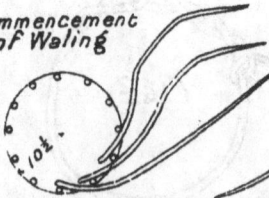
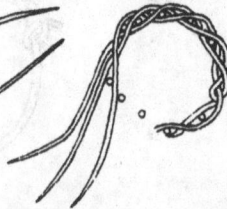
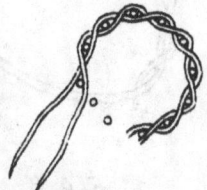

Fig. 4.
BRUSHWOOD GABION.
Sewing and Pairing rods not shown

←---2.0 ---→

2'9"

Fig 5.
EXPANDED METAL GABION.

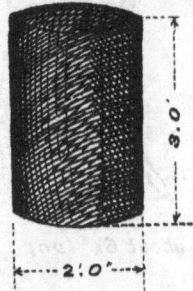

3'0"

←--- 2'0' ---→

PAPER BAND GABION.

Fig 6.

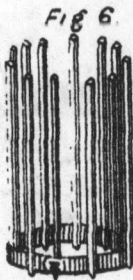

Fig. 7

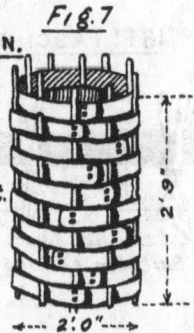

All joints in the bands to be in the same half of the circumference and buried in the parapet.

2'9"

←--- 2'0" ---→

Malby & Sons Lith

Plate 5

HURDLES.

Fig. I.

Commencement of a 6 ft Hurdle

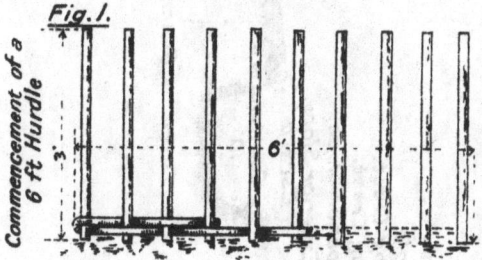

3' 6'

Fig 2.

1 to 2 Bundles Brushwood Weight 56 lbs.

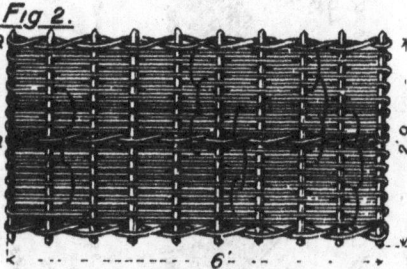

3 Men, 2½ hours, Tools, 2 Bill Hooks, 2 Gabion Knives, 1 Mallet. One 5 ft Rod. 1 Pliers. (if sewn with wire). 1 Gauge (made on ground.)

a a

a a

6' 2.9

a - pairing rods

ROUGH HURDLES.

Fig.3.

← - - - - - - - - 6' 0" - - - - - - - →

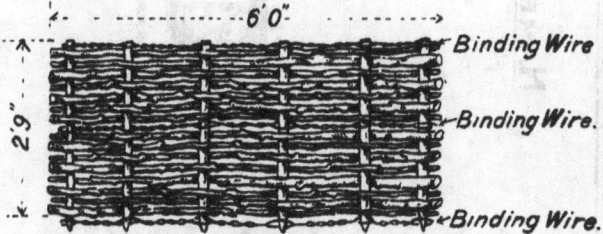

→ Binding Wire

→ Binding Wire.

→ Binding Wire.

2' 9"

ELEVATION.

PLAN.
(Shewing pickets and binding wire.)

Malby & Sons. Lith

Plate 6

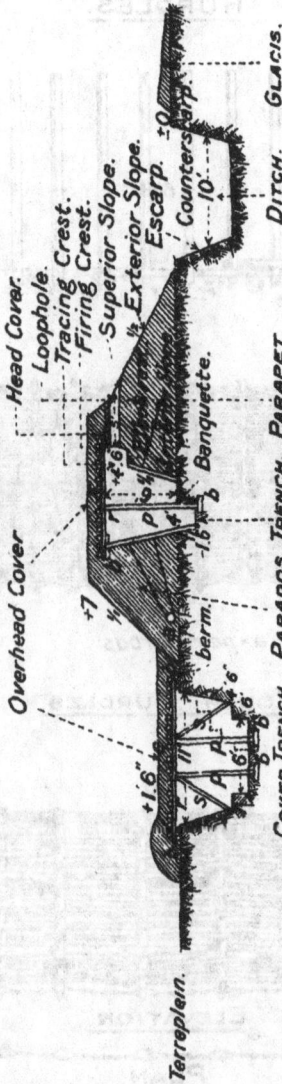

NAMES OF PARTS OF A FIELD WORK.

(SEE ALSO PLATE 21).

Head Cover.
Loophole.
Tracing Crest.
Firing Crest.
Superior Slope.
Exterior Slope.
Escarp.
Counterscarp.

Overhead Cover

Loophole.

Banquette.

Terreplain.

COVER TRENCH. PARADOS. TRENCH. PARAPET. DITCH. GLACIS.

p = posts
s = struts.
r = rafters.
b = bearing plates.
c = covering materials.
a = anchorage to revetments.

CUTTINGS & EMBANKMENTS.

EMBANKMENTS.

Fig. 1.

Recesses as required

Continuous
Communication Path as close behind banquette as possible.

Fig. 2.

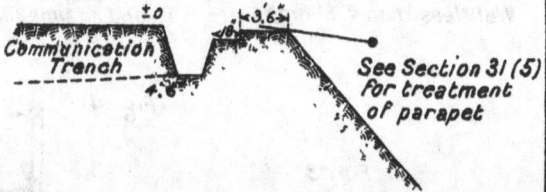

Communication Trench

See Section 31 (5)
for treatment
of parapet

RAILWAY CUTTING

Fig 3

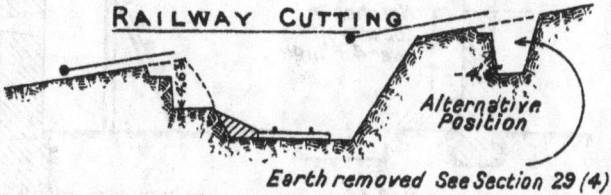

Alternative
Position

Earth removed See Section 29 (4)

Fig. 4

HEDGES.

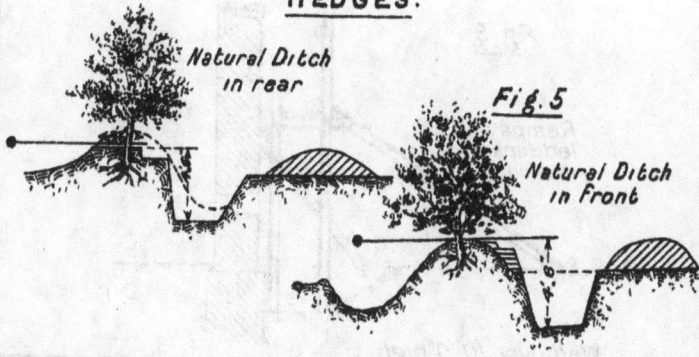

Natural Ditch
in rear

Fig. 5

Natural Ditch
in front

Plate 8

WALLS.

[For loopholes see Plate 10]

Fig 1

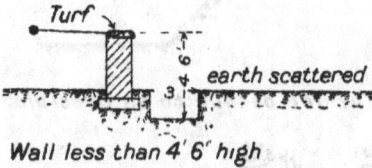

Turf

earth scattered

Wall less than 4' 6" high

Fig 2

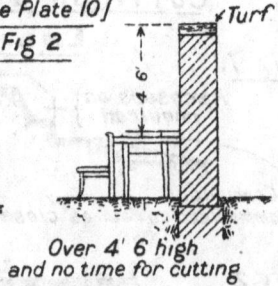

Turf

4' 6"

Over 4' 6" high
and no time for cutting

Fig. 3

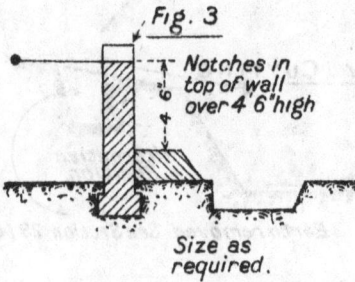

Turf

Notches in
top of wall
over 4' 6" high

4' 6"

Size as
required.

Fig 4

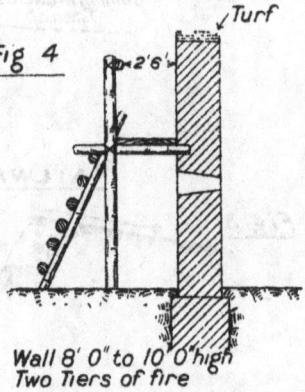

Turf

2' 6"

Wall 8' 0" to 10' 0" high
Two Tiers of fire

Fig 5

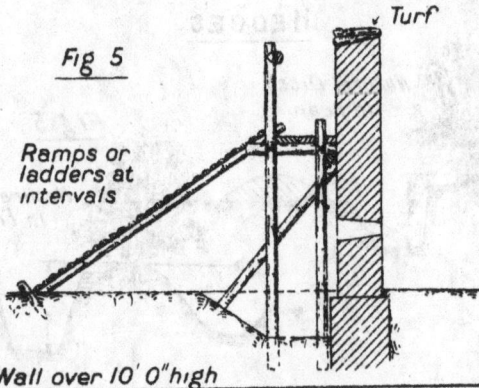

Turf

Ramps or
ladders at
intervals

Wall over 10' 0" high

Malby & Sons. Lith

Plate 9

FIRE TRENCHES.

Fig. 1.

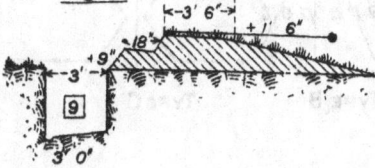

Fig. 2

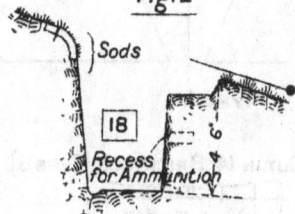

}Sods

Recess for Ammunition

Fig. 3.

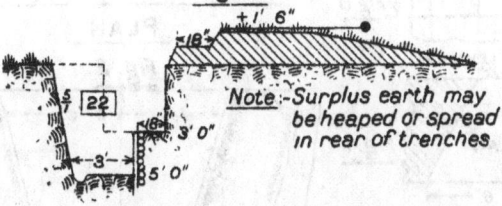

Note:- Surplus earth may be heaped or spread in rear of trenches

COVER — LYING DOWN.

Fig. 4.

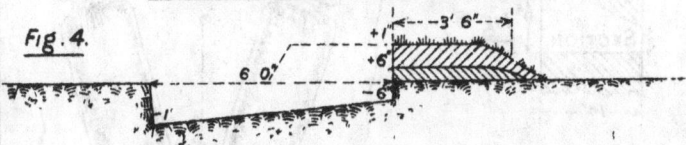

Fig. 5

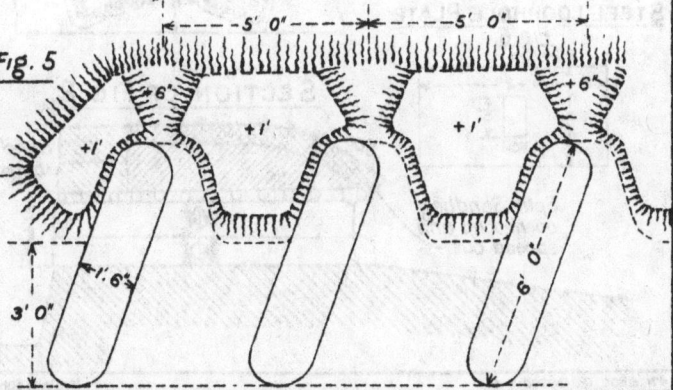

Plate 10

LOOPHOLES.

Fig. 1.

para pet

TYPE A   TYPE B   TYPE C

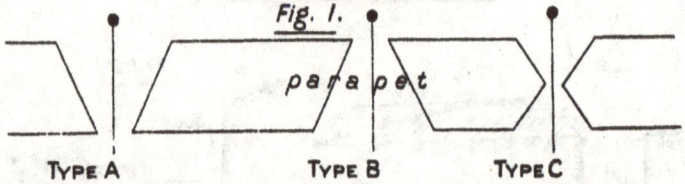

CUT IN 14" BRICKWALL [TYPE B]

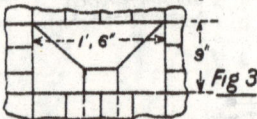

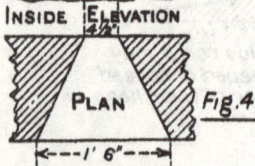

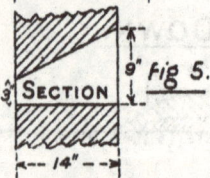

1', 6"   9"

Fig 3

INSIDE ELEVATION   4½"

PLAN   Fig. 4.

1' 6"

SECTION   Fig 5.   9"

14"

SANDBAG LOOPHOLE

PLAN

Fig. 2.

2' 6"

3' 6"

6"

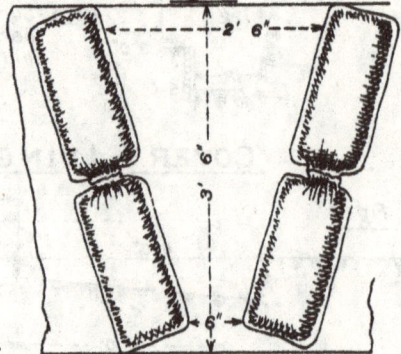

STEEL LOOPHOLE PLATE

Fig. 6.

9/16 thick

2'

5"

3"

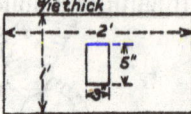

SECTION OF FIG. 2.

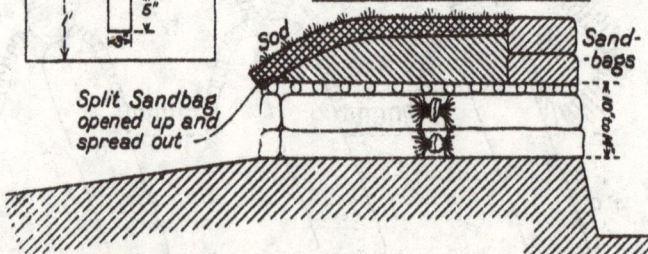

Sod   Sand-bags

Split Sandbag
opened up and
spread out

Plate 11

LOOPHOLES.
SHINGLE LOOPHOLE (TYPE A).
PLAN AT FIRING LEVEL.

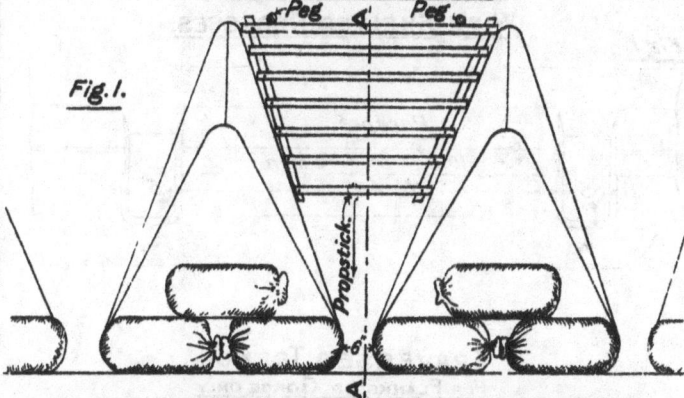

Fig.1.

Peg   Peg

Propstick.

6"

SECTION OF FIG. I, AT A.A.

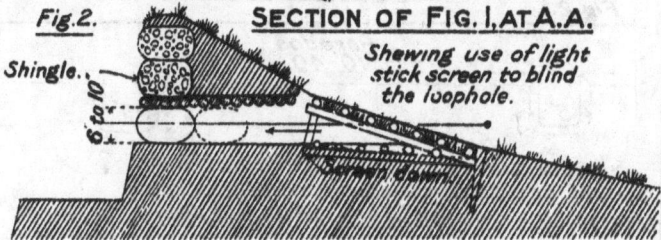

Fig.2.

Shingle.

6 to 10

Shewing use of light stick screen to blind the loophole.

SHINGLE LOOPHOLE (TYPE C).

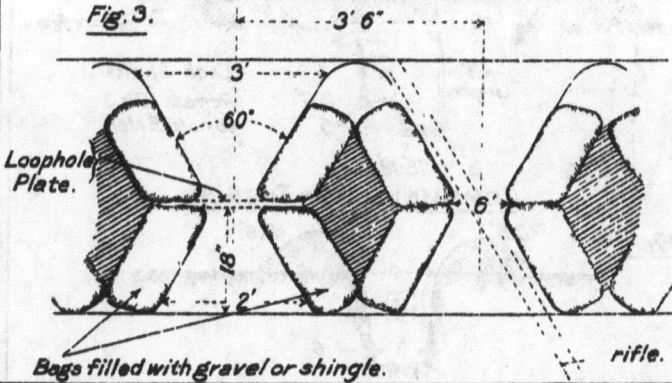

Fig.3.

3' 6"

3'

60°

6'

Loophole Plate.

18"

2'

Bags filled with gravel or shingle.

rifle.

Plate 12

TRAVERSED TRENCH.
WITH SHELTER RECESSES.

Fig I.

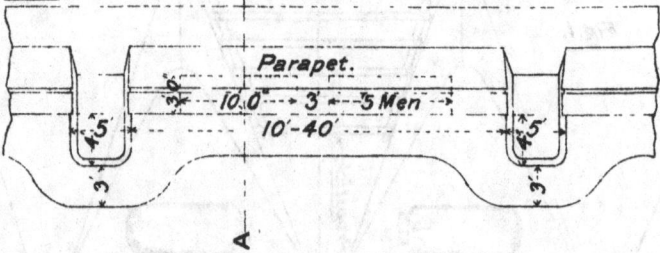

TRAVERSED TRENCH.
FOR FLANKS OR GORGE ONLY

Fig.2.

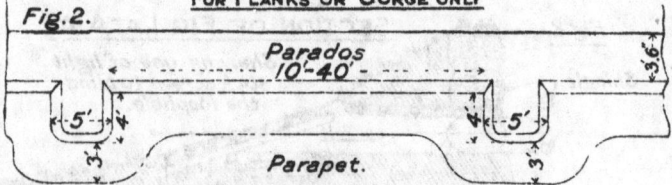

SECTION ON A.A. FIG. I.

Fig.3.

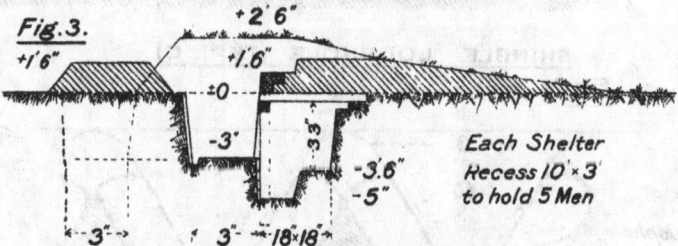

Each Shelter Recess 10' × 3' to hold 5 Men

COMMUNICATION TRENCH.

Fig.4

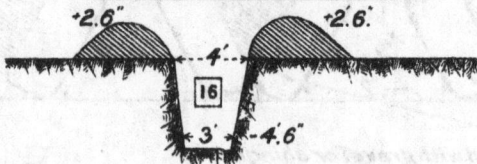

Plate 13

TRAVERSED AND RECESSED TRENCH

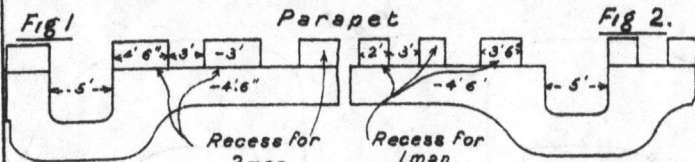

Fig 1. Parapet Fig 2.

4'6" 3' 3' 2' 3' 3'6'
5' 4'6" 4'6" 5'

Recess for 2 men

Recess for 1 man

2' to 3'6" wide according to angle of view required. Roofing material 4 ft long will do if it is only required to roof recesses 2 ft wide

SECTION THRO RECESS.

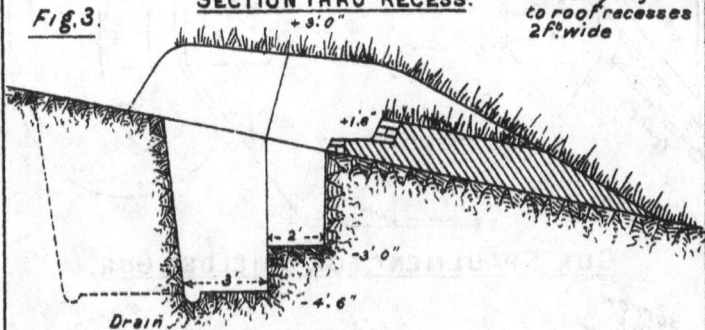

Fig. 3.

+3'0"

+1'6"

2'
3'0"
3'
4'6"

Drain

OVERHEAD COVER.
Added to Fig 3.

Fig. 4.

4'6"

2'6" +1'6"

5'6"

2'
3'0"
3'
4'6"

Steps at intervals

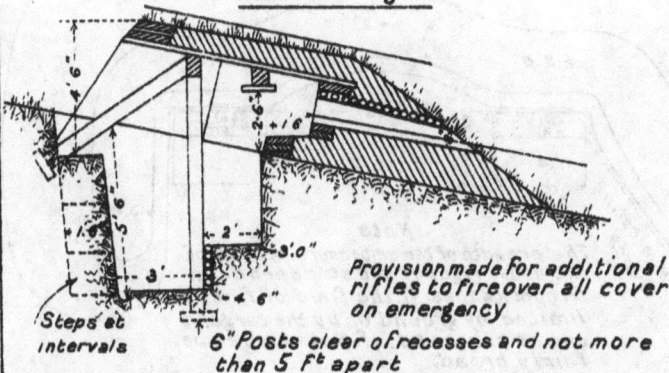

Provision made for additional rifles to fire over all cover on emergency

6" Posts clear of recesses and not more than 5 ft apart

471 8699/183 50000 7 11   Malby & Sons, Lith

Plate 14

GUN PIT FOR SHIELDED GUN.

Fig. 1.

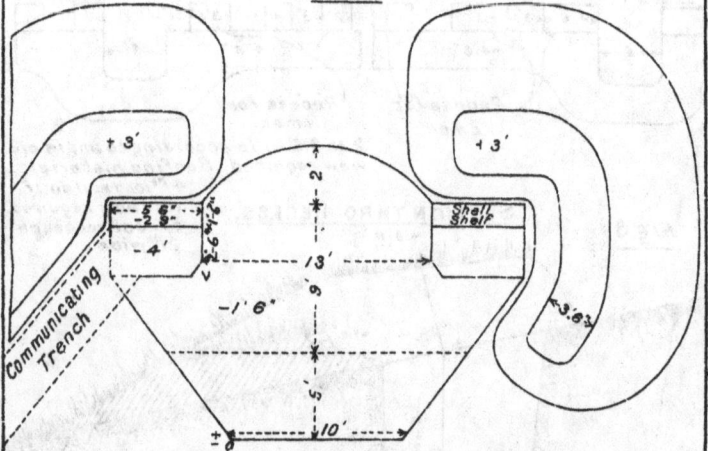

GUN EPAULMENT FOR SHIELDED GUN.

Fig 2

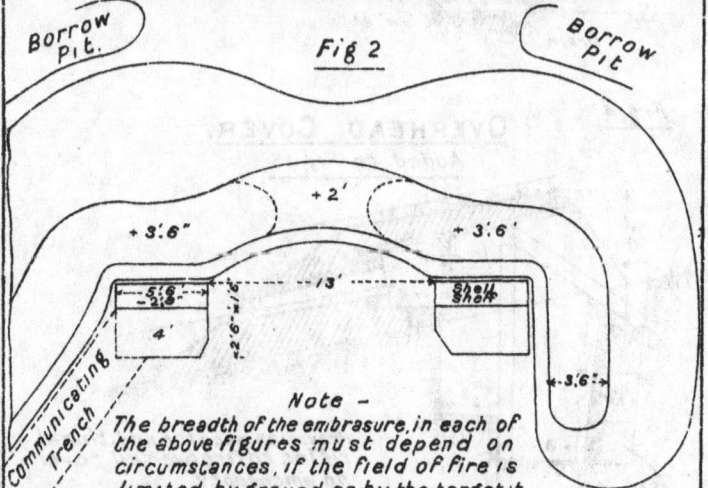

Note –
The breadth of the embrasure, in each of the above figures must depend on circumstances, if the field of fire is limited by ground or by the target it can be narrowed, if not it must be fairly broad.

Malby & Sons. Lith

Plate 15

MACHINE GUN EMPLACEMENT.

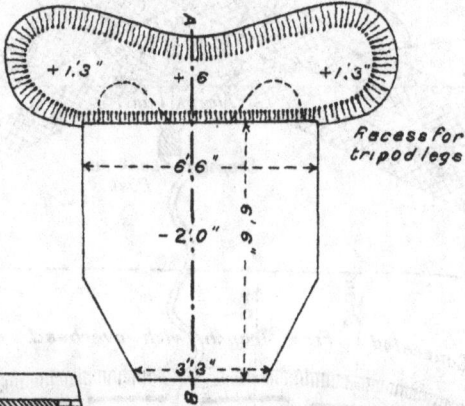

+1.3"   +.6"   +1.3"

Recess for
tripod legs

6.6"

9.9'

Fig 1
PLAN.

-2.0"

3.3"

+1.3"

6.6"

SECTION ON A.B.

Fig 2.
PLAN
Showing supports
of head cover

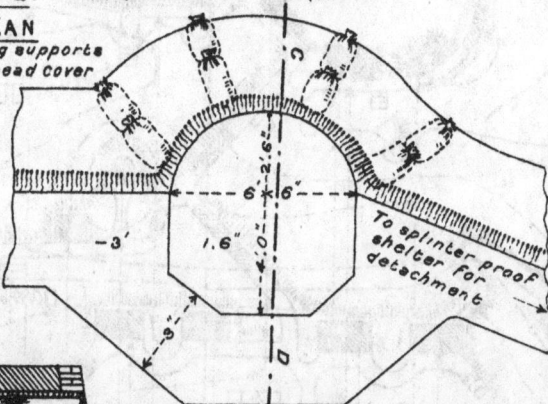

C

6" 6"

2.6"

-3'

1.6"

4.0"

3"

To splinter proof
shelter for
detachment

D

+2

Height to be regulated by trial with the gun
sights

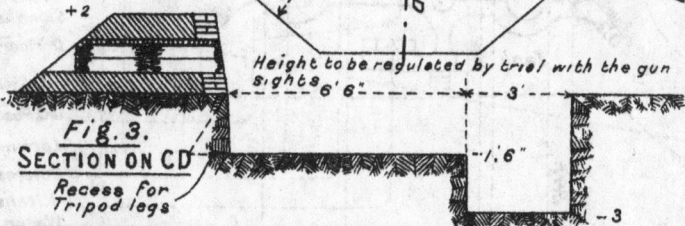

6.6"   3'

Fig. 3.
SECTION ON CD
Recess for
Tripod legs

-1.6"

-3

DIAGRAM.

Plate16

To Illustrate conventionally the Auxiliary Works required to prepare a Fire Trench for occupation and use (not to any Scale)

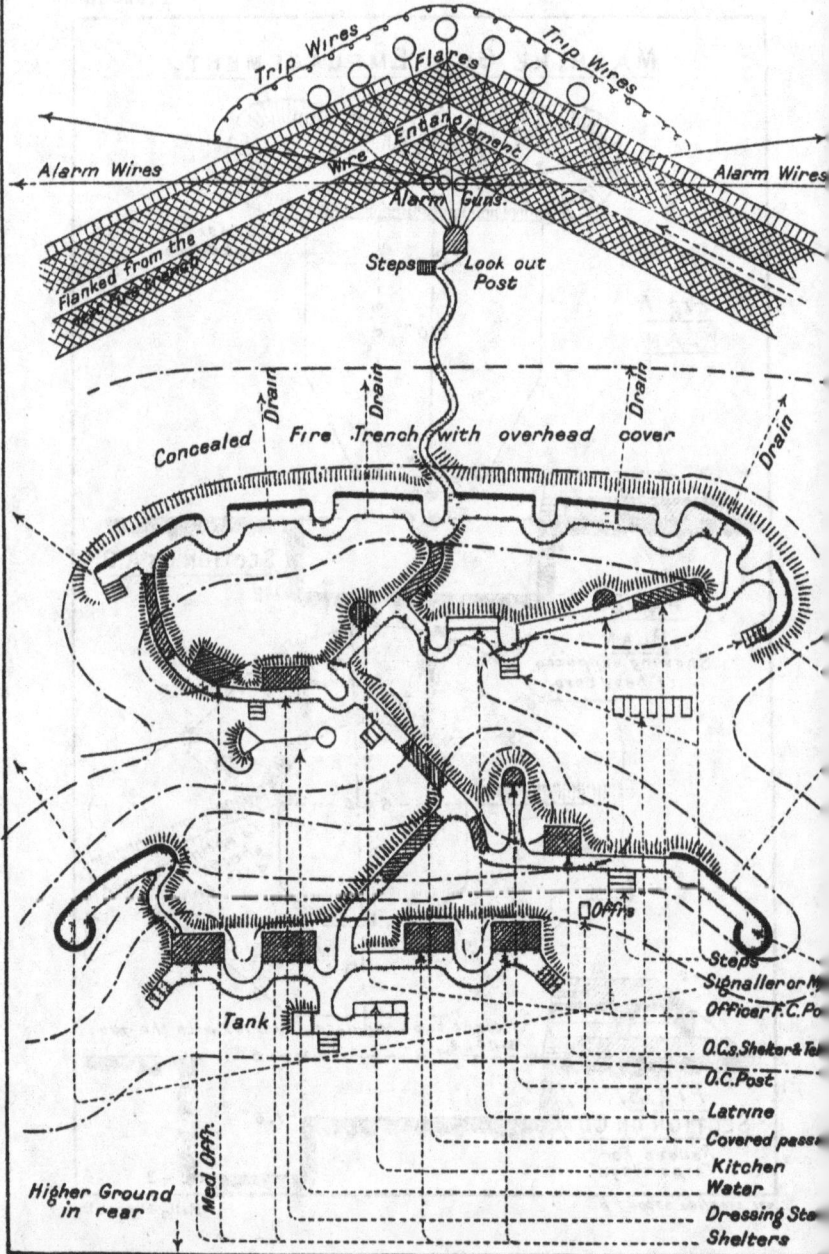

Trip Wires

Flares

Trip Wires

Alarm Wires

Wire Entanglement

Alarm Wires

Alarm Guns.

Flanked from the next Fire Trench

Steps

Look out Post

Drain

Drain

Drain

Drain

Concealed Fire Trench with overhead cover

Steps

Signaller or M.

Officer F.C. Po.

O.C.s Shelter & Tel.

O.C. Post

Latrine

Covered passa.

Kitchen

Water

Dressing Sta.

Shelters

Tank

Offic.

Med. Offr.

Higher Ground in rear

471.8609/189 30000.7 II

Plate 18

DIAGRAM. A

To illustrate defensive co-operation between 'Closed Groups', intermediate, and Dummy, Groups of Fire Trenches; strengthened with Obstacles and Alarm Guns, and supported by Machine Guns, Field Guns and Howitzers. The dispositions are diagrammatic, not tactical.

Not to Scale.

The reference numbers are those of the Sections in Manual. Field Telephones and Search lights are not illustrated.

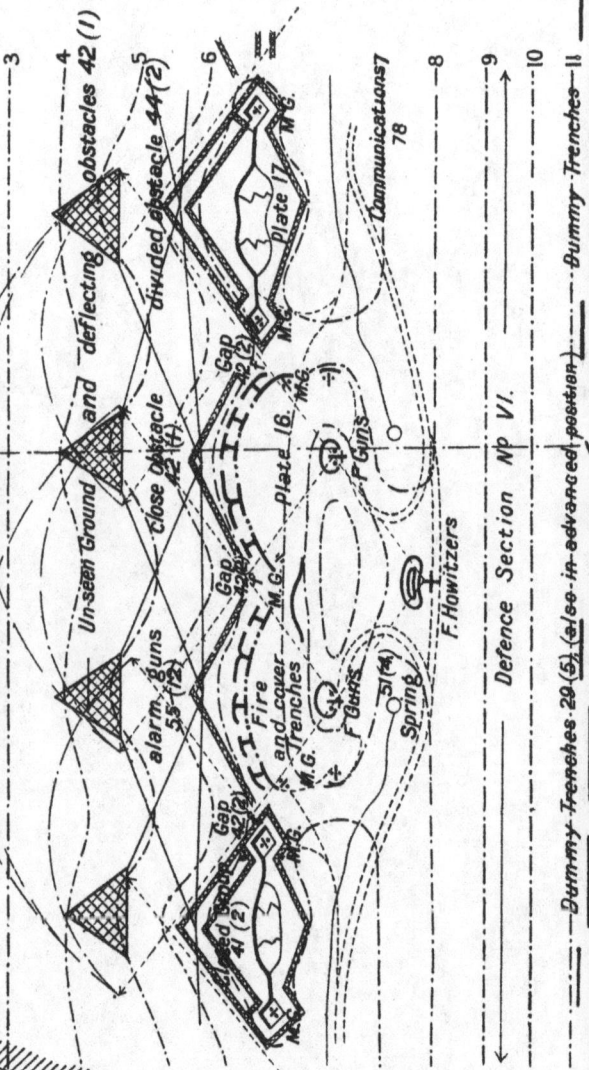

obstacles 42 (1)

divided obstacle 44 (2)

Un-seen Ground and deflecting

Close Obstacle 42 (1)

alarm guns 53 (12)

Gap 42 (2)

Gap 42 (2)

Gap 42 (2)

Fire and cover Trenches

Gap 42 (2)

Fixed Dummy 41 (2)

M.G.

M.G.

M.G.

M.G.

M.G.

Plate 17

Plate 16

F. Guns

51 (4)

Spring

F. Guns

F. Howitzers

Communications 7

78

Defence Section No VI.

Dummy-Trenches 29 (5) (also in advanced position)

Dummy Trenches 11

SECTION ON A.A.

2
3
4
5
6
7
8
9
10
11
12

471 8609/163 30000 7 11

Malby & Sons.

Plate 19.

LOW COMMAND REDOUBT.

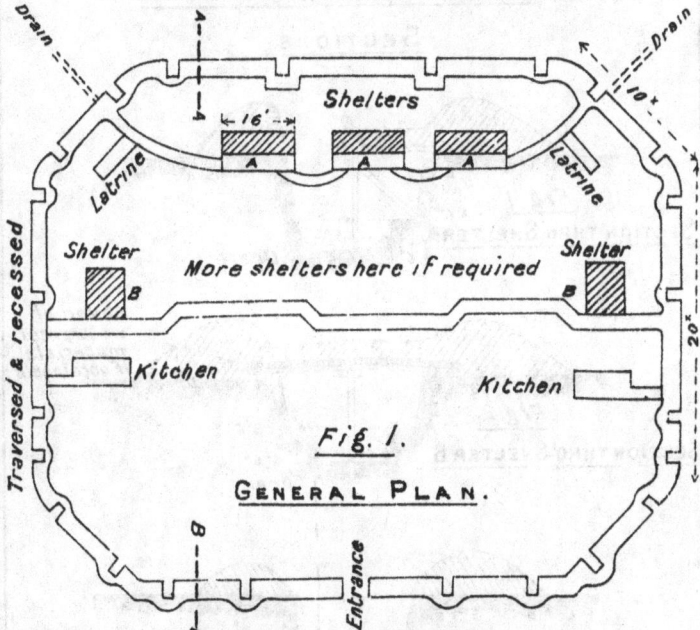

Drain

Shelters

16'

A · A · A

Latrine

Latrine

Traversed & recessed

Shelter
B

More shelters here if required

Shelter
B

Drain

10 x

20 x

Kitchen

Kitchen

Fig. 1.

GENERAL PLAN.

B

B

Entrance

SECTION THRO FACE A.A.

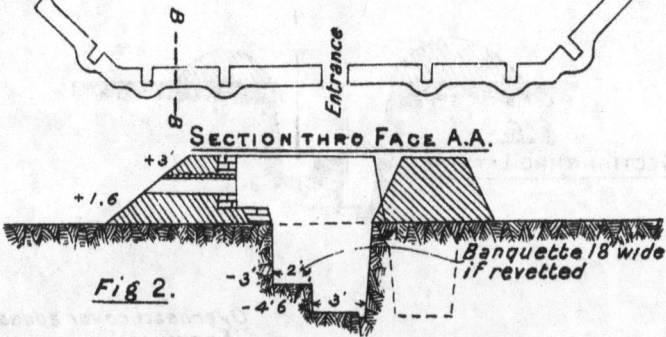

+3'

+1.6

-3' 2'

-4'6

3'

Banquette 18' wide
if revetted

Fig. 2.

SECTION THRO GORGE B.B.

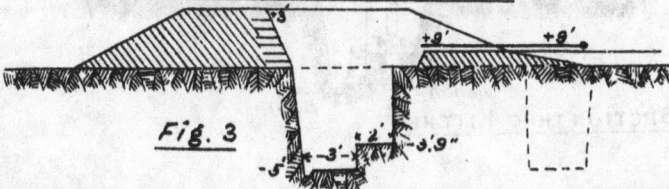

3'

+8'

+8'

-5' -3' 2' -3'.9"

Fig. 3

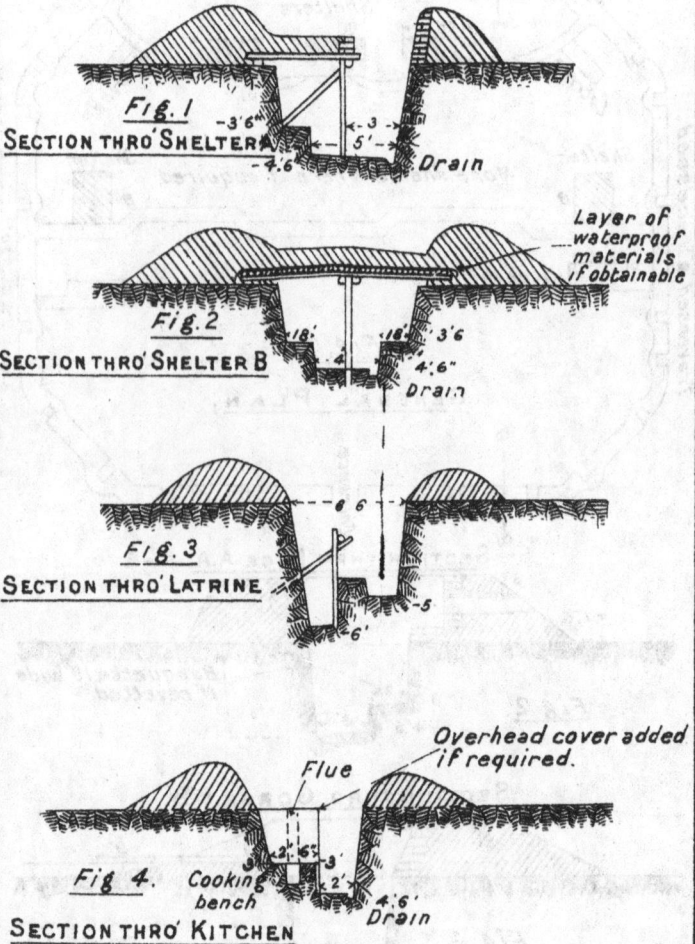

Plate 20

LOW COMMAND REDOUBT.

SECTIONS

Fig. 1
SECTION THRO' SHELTER A
−3'6" ←3→ 5' −4'6" Drain

Fig. 2
SECTION THRO' SHELTER B
Layer of waterproof materials if obtainable
18" 4' 18" 3'6" 4'6" Drain

Fig. 3
SECTION THRO' LATRINE
6'6" 6' 5'

Fig. 4. Cooking bench
SECTION THRO' KITCHEN
Flue
Overhead cover added if required.
2'6" 3' 2' 4'6" Drain

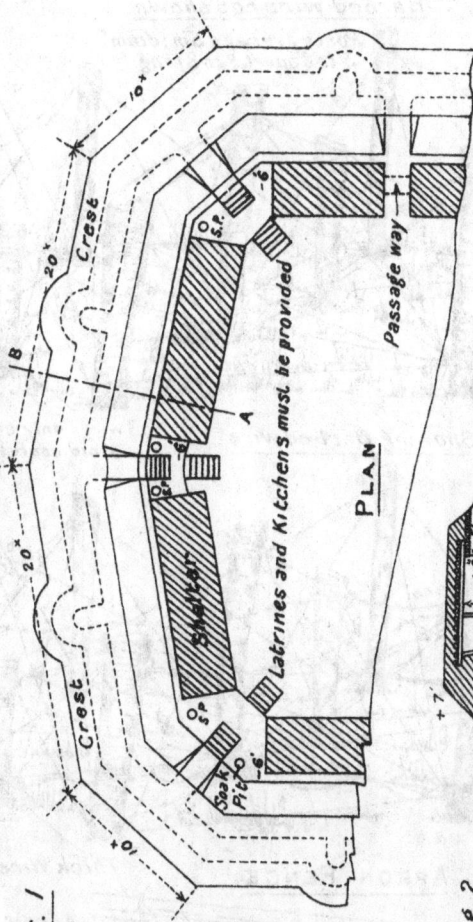

HIGH COMMAND REDOUBT.

Plate 21

Fig. 1

PLAN

Latrines and Kitchens must be provided

Passage way

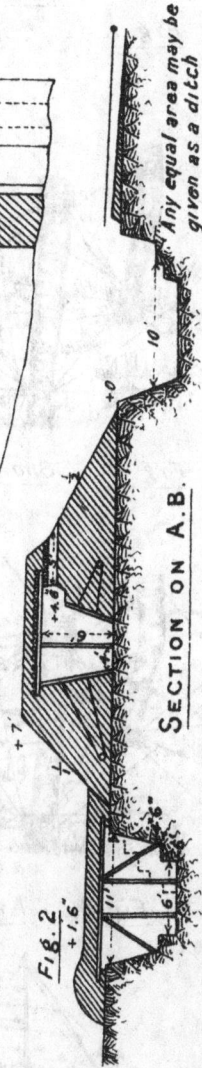

SECTION ON A.B.

Fig. 2

Any equal area may be given as a ditch

Malby & Sons, Lith

Plate 22

HIGH WIRE ENTANGLEMENT.

Fig. 1. Barbed wire not shown.

Posts average 5in: diam^r.
5' to 8' apart, 5' to 8' long

Fig. 2. Showing Barbed wire   (3 wires only shown to avoid confusion)

Barbed   wire

Barbed wire

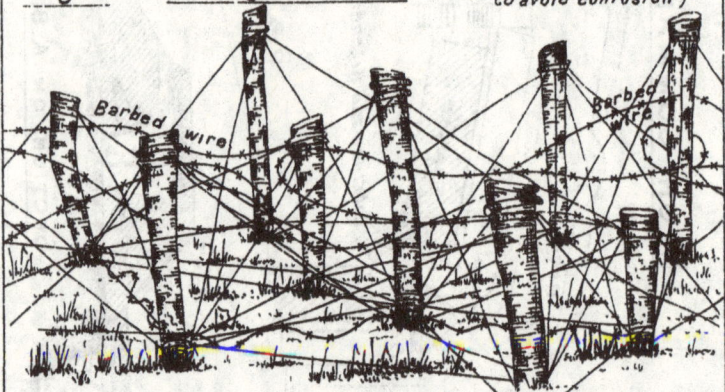

Fig. 3. APRON FENCE.   Thick Wire

Apron

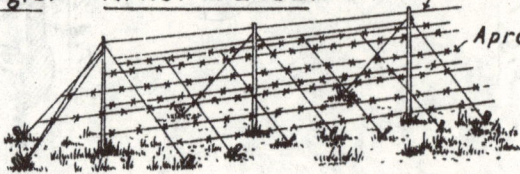

Malby & Sons, Lith.

Plate 28.

DEFENCE OF A HOUSE

DOOR.

Fig. 1.

EXTERNAL ELEVATION

Planks nailed to wall Gravel filled in between planks and door

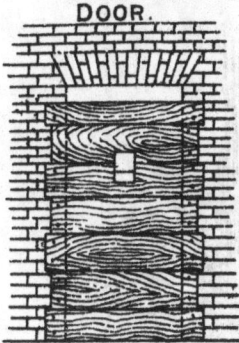

Fig. 2. SECTION

A

Shingle or Gravel

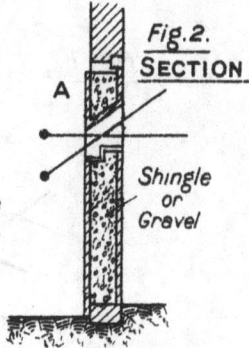

WINDOW.

Fig. 3.

INTERNAL ELEVATION

Sacks filled with Shingle or Gravel

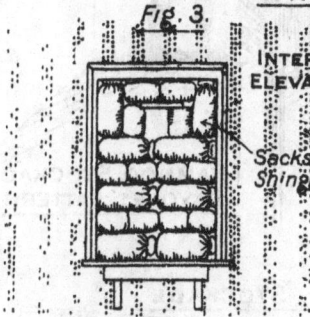

Fig 4. SECTION

Glass removed

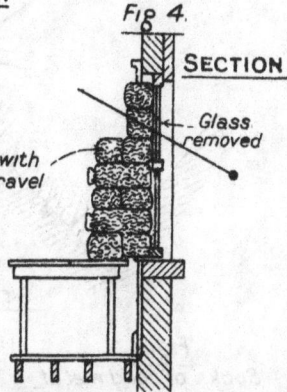

SECTION.

Fig. 4

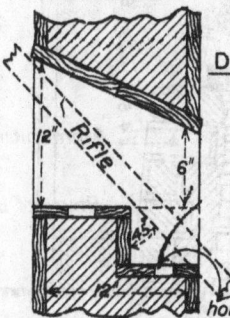

12"   Rifle   6"

45°

DETAILS OF LOOPHOLE AT A IN FIG. 2.

PLAN

90°

12"   3"

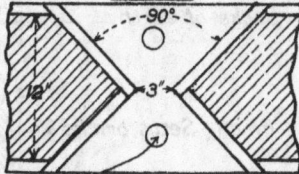

holes to refill with shingle after settlement

Plate 24

STOCKADES.

Fig. 1

4' 0"

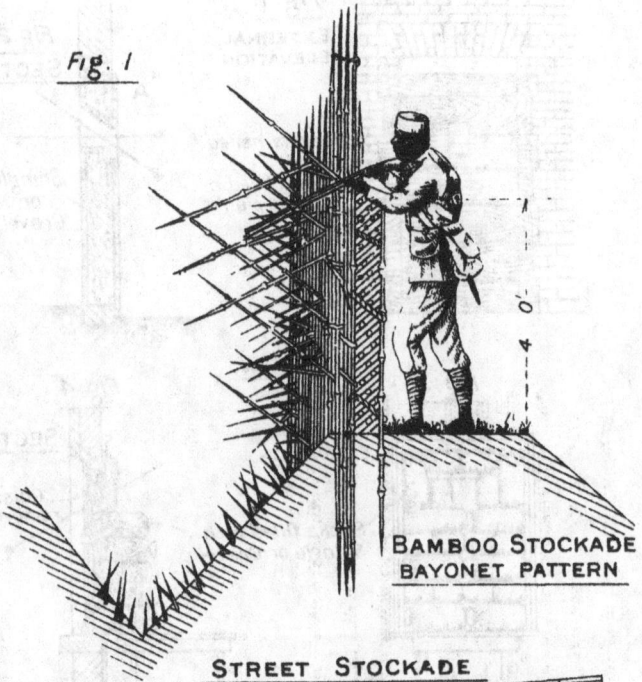

BAMBOO STOCKADE
BAYONET PATTERN

STREET STOCKADE

Fig. 2
Sacks of road metal

6'

2' 6"

6'

Sacks of earth

4' 6"

Ammunition

6' 0"

Boxes of Earth

Paving Setts, bricks &c

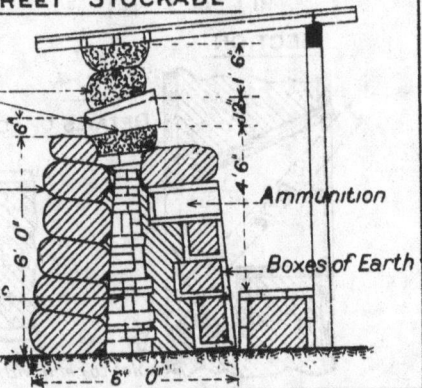

5' 0"

Plate 25.

STOCKADES

Fig 1.
TIMBER AND SHINGLE.

+6'.6"

Uprights about 4' apart lashed to pickets driven into ground

+3'.9" Elbowrest
Planking corrugated Iron etc.
6' Shingle

Loopholes
3' to 5' apart

Wire obstacle

Fig 3.
cleared

Cleared

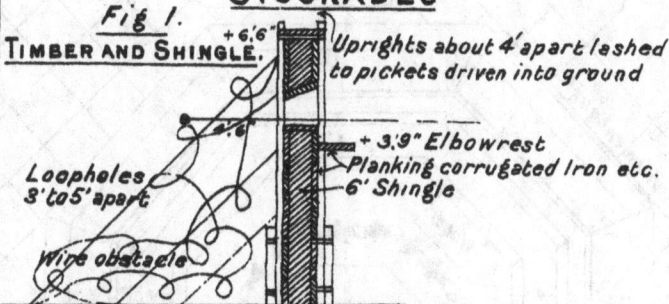

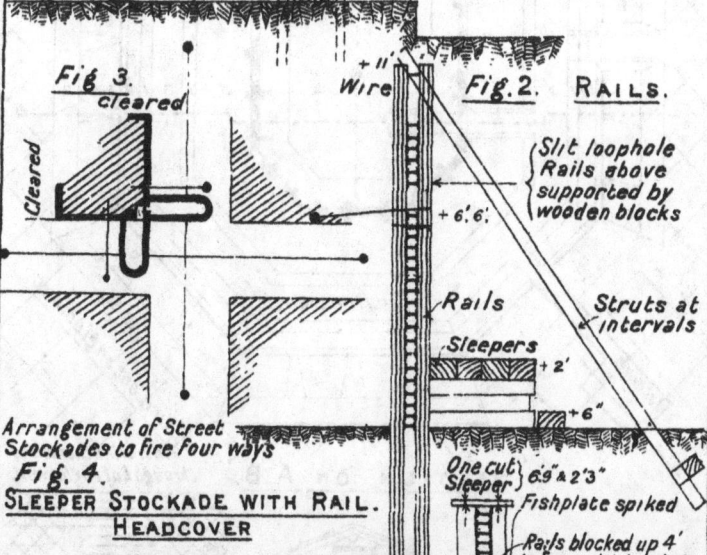

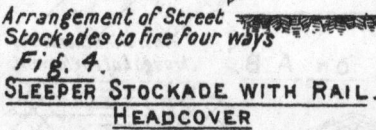

Arrangement of Street
Stockades to fire four ways

Fig. 4.
SLEEPER STOCKADE WITH RAIL.
HEADCOVER

Showing Earth or Coal
parapet

Earth—

Wire

+11"

Fig. 2.   RAILS.

Slit loophole
Rails above
supported by
wooden blocks

+6'.6"

Rails

Sleepers   +2'

+6'

Struts at
intervals

One cut
Sleeper   6'.9" & 2'.3"
Fishplate spiked

Rails blocked up 4'
to form slit loophole

3'.6"   1'.6"   Elbow Rest

Turf

Stone
or
Coal
18"   3'.9"   4'.6"

6'

Borrow
Trench

Malby & Sons, Lith.

Plate 26

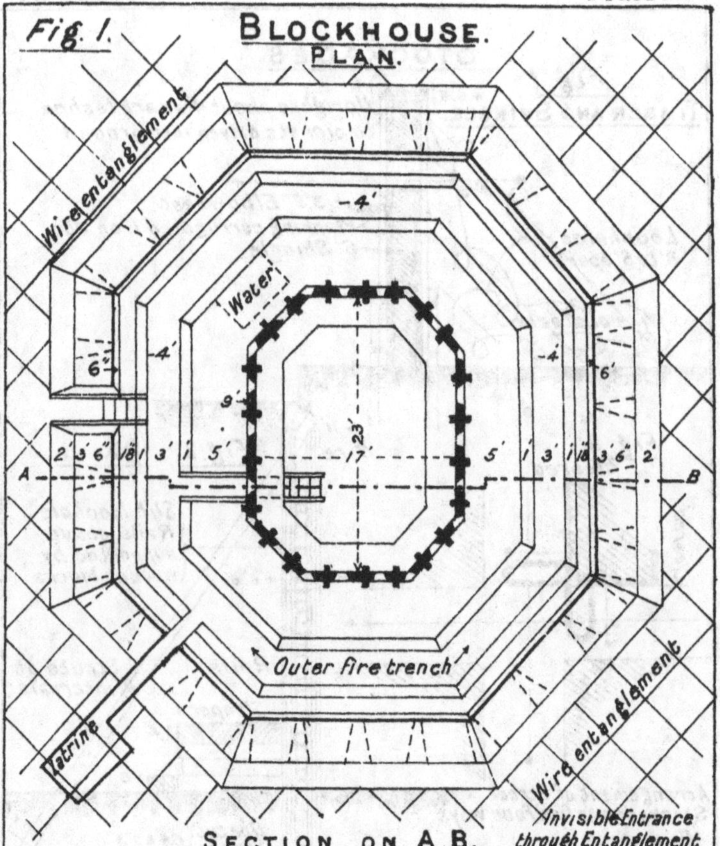

Fig. I.

BLOCKHOUSE.
PLAN.

Wire entanglement

Water

-4'

6'

4'

-4'

6'

9'

23'

2' 3' 6' 18 3' 5' 17 5' 1' 3' 18 3 6' 2'

A ————————————————————————— B

Outer fire trench

Latrine

Wire entanglement

SECTION ON A.B.

Invisible Entrance through Entanglement

Fig. 2

3' 6' +3'
18" 5' 5' 6.6 +2' 5' 5' +3' 3' 6'
4' 3' 4' 3' 4'

8' of broken stone

Fig 3.

471 8668/185 30600 7 11

Malby & Sons, Lith

Plate 27

DEFENSIBLE POST.

PLAN.

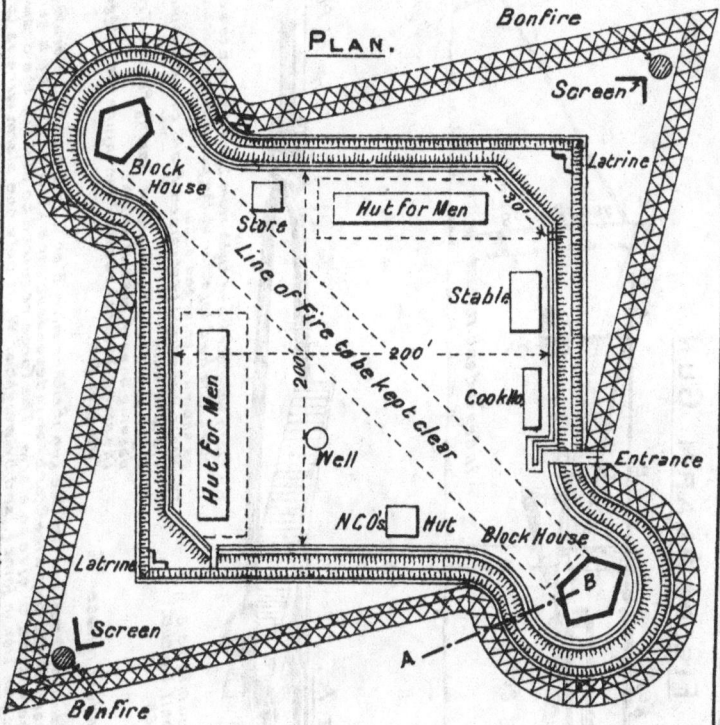

Bonfire

Screen

Block House

Latrine

Store

Hut for Men

30'

Stable

Line of Fire to be kept clear

200'

Cook Ho.

200'

Well

Entrance

Hut for Men

Latrine

NCOs Hut

Block House

Screen

B

Bonfire

A

SECTION ON A.B.

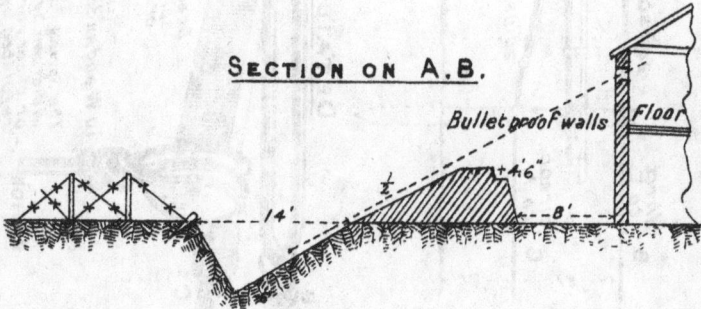

Bullet proof walls

Floor

14'

½

+4'6"

8'

Plate 28

FIELD ALARM GUN.

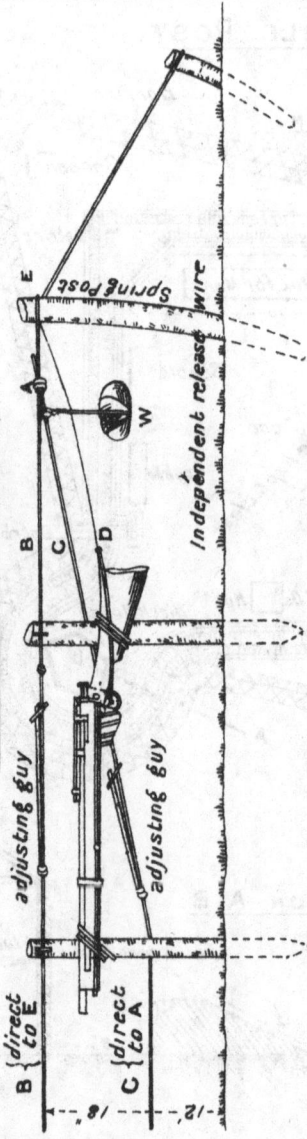

B { direct to E

C { direct to A

adjusting guy

adjusting guy

Spring Post

Independent release wire

W

Independent release wire

To W and independent release

DETAIL AT A.

To Spring Post E

Draw-pin may be a spiked nail, bayonet or stick.

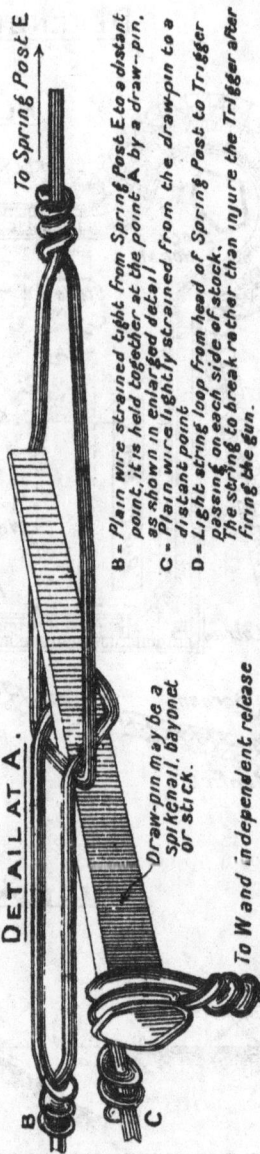

B = Plain wire strained tight from Spring Post E to a distant point, it is held together at the point A by a draw-pin, as shown in enlarged detail.

C = Plain wire lightly strained from the draw-pin to a distant point

D = Light string loop from head of Spring Post to Trigger passing on each side of stock. The string to break rather than injure the Trigger after firing the gun.

ACTION :- The Spring Post will fly back and fire the gun if either wire B or C is cut or if the draw-pin is withdrawn by means of the trip wire C; or by an independent wire at the will of a sentry or observer who wishes to give an alarm. The force required to pull out the draw-pin is regulated by its position in joint A, and by weight of W. If only one alarm wire is to be run out use wire C. Wire B can be made fast to any fixed point near the gun.

B

C

Plate 29

CARTRIDGE ALARM.
WITH FLARE.
GENERAL VIEW

Fig 1.

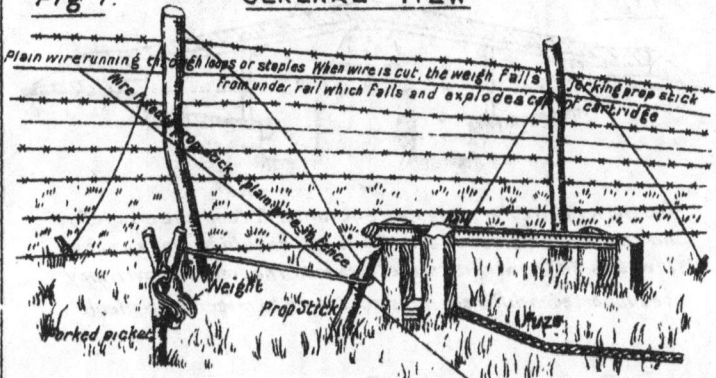

Plain wire running through loops or staples When wire is cut, the weight falls from under rail which falls and explodes cap of cartridge

Locking prop stick

Wire leading to prop stick along wire netting fence

Weight

Prop Stick

Fuze

Forked picket

Fig 2.

PLAN

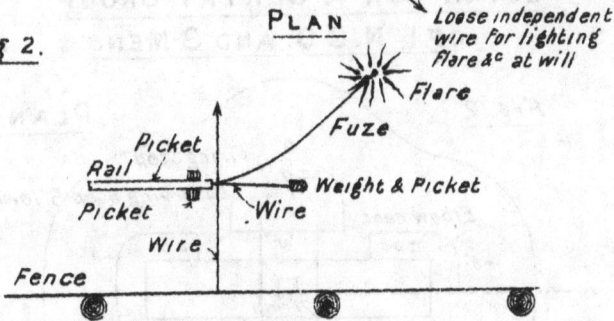

Loose independent wire for lighting flare &c at will

Flare

Fuze

Picket

Rail

Picket

Weight & Picket

Wire

Wire

Fence

Fig 3

ELEVATION.

Two stout pickets acting as guides to rail when falling

Piece of rail which falls on spike when prop-stick is jerked away, detonating cap of cartridge

Spike

Piece of wood with hole through centre to keep spike erect.

Piece of hard wood with a piece of tin nailed on top A blank cartridge is passed through a hole in its centre the rim of cartridge resting on the tin.

Plate 30

COVER FOR AN OUTPOST PIQUET.

Fig 1

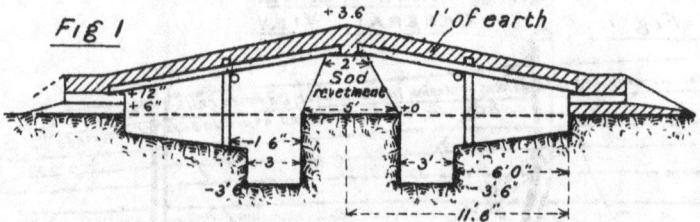

The work is circular in plan. An entrance 3 feet wide is left on the least exposed face. Drainage should be made on the lowest side. The work is covered with sods. The overhead cover may be supported in different ways according to material available

COVER FOR A SENTRY GROUP.
OF 1 N.C.O. AND 3 MEN.

Fig. 2.   PLAN.

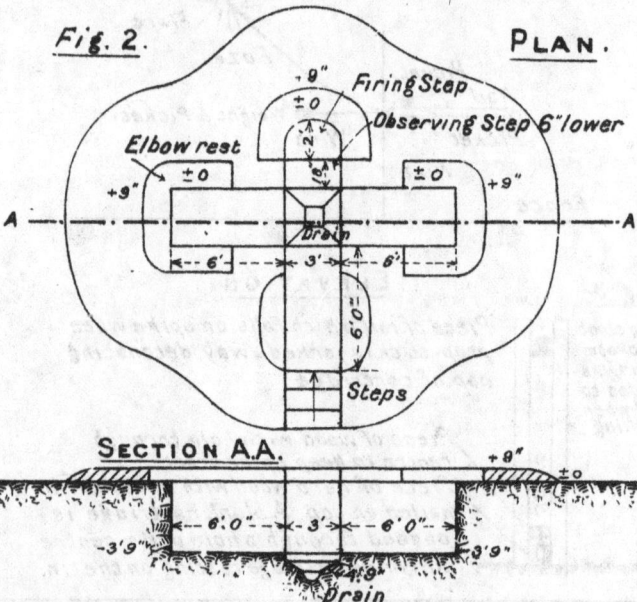

Firing Step

Observing Step 6" lower

Elbow rest

Drain

Steps

SECTION A.A.

Drain

Plate 31

FIELD GEOMETRY.

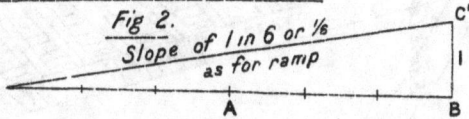

Fig 2.
Slope of 1 in 6 or ⅙
as for ramp

Fig 1.
Slope of 4 in 1 or 4/1 as for Revetments

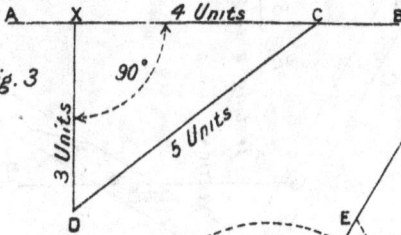

Fig. 3
A X 4 Units C B
90°
3 Units
5 Units
D

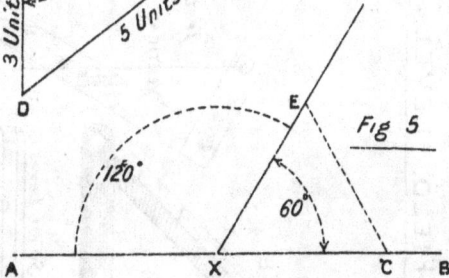

Fig 5
120°
60°
A X C B
E

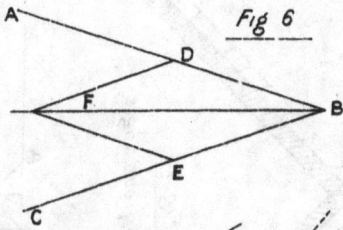

Fig 6
A D B
F
C E

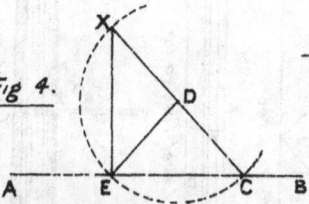

Fig 4.
X D
A E C B

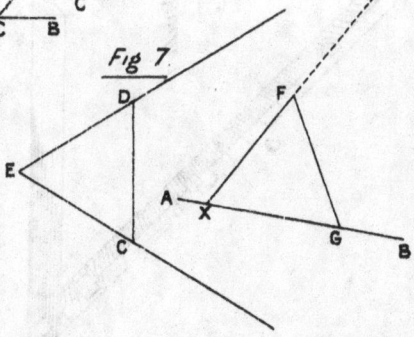

Fig 7
D
E
C
A X G B
F

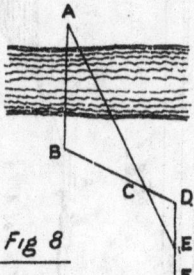

Fig 8
A
B C D
E
F

Plate 32

FIELD LEVEL

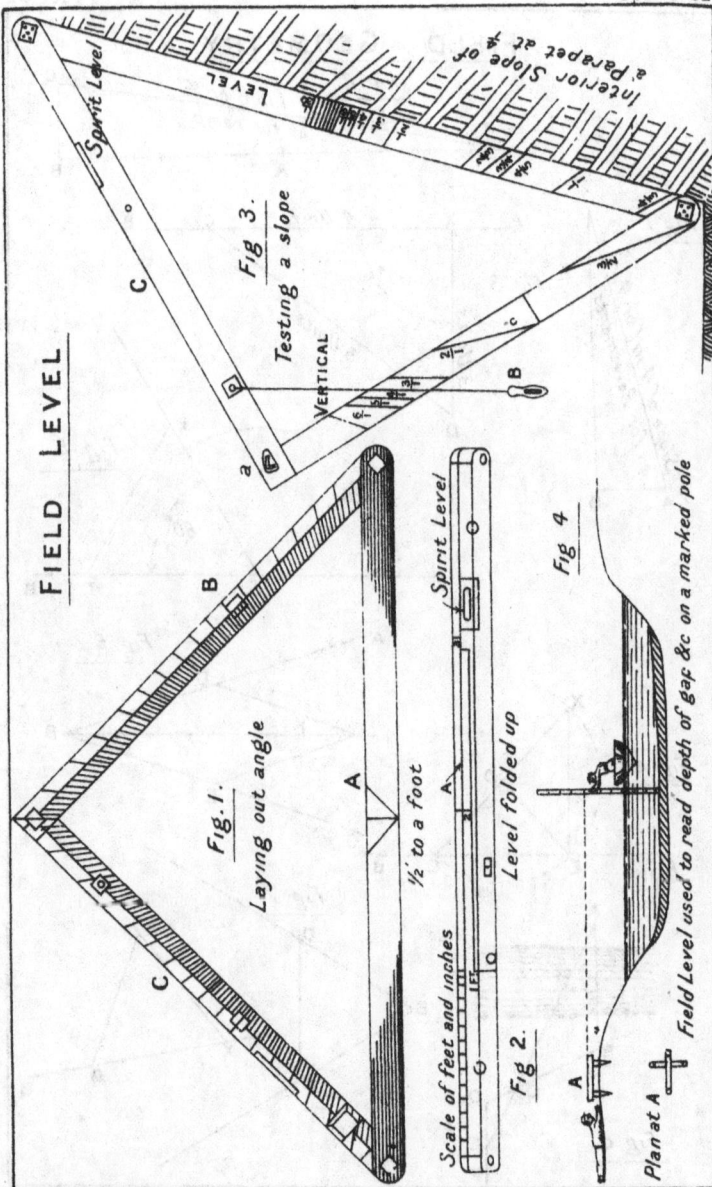

Fig. I.
Laying out angle

Scale of feet and inches

½ to a foot

Spirit Level

Level folded up

Fig. 2.

Fig 3.

Testing a slope

Interior Slope of a parapet at ¼ slope

LEVEL

Spirit Level

VERTICAL

Fig 4

Plan at A

Field Level used to read depth of gap &c on a marked pole

471 8683/183 30000 7 11

Malby & Sons, Lith

Plate 33

COOKING IN THE FIELD.

Fig 1

"Kettles, Camp Oval, 12 quarts", 13½ in.× 9 in. ×
11 in high

Fig. 2.

Malby & Sons, Lith

Plate 34.

KITCHENS.

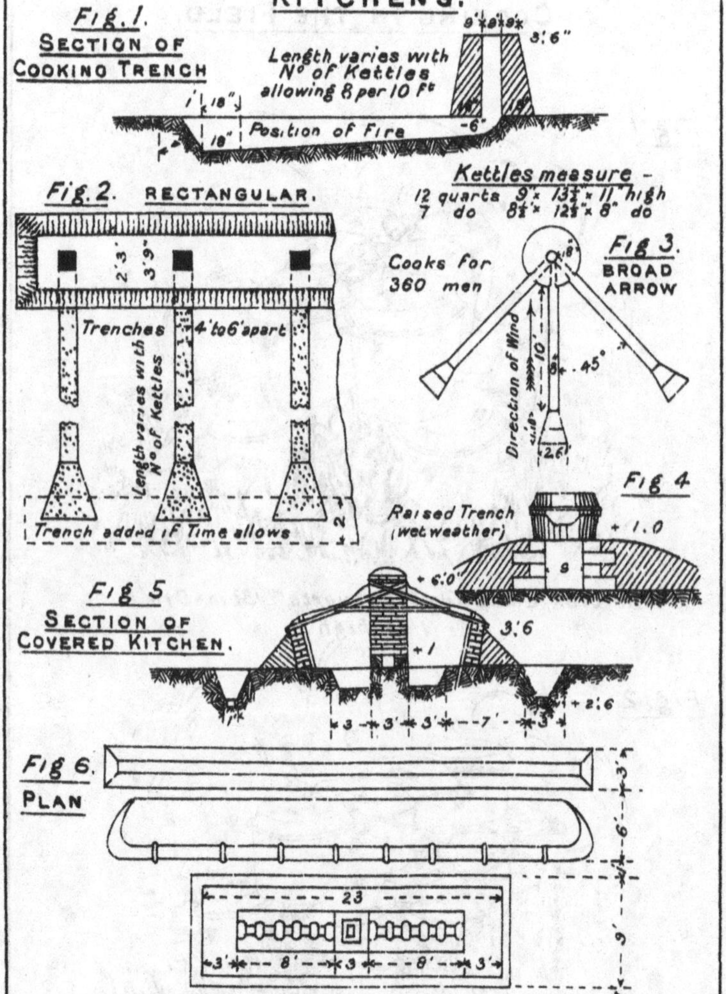

Fig. 1.
SECTION OF COOKING TRENCH

Length varies with Nº of Kettles allowing 8 per 10 Fᵗ

3' 6"

1' 18"

Position of Fire

18"

6"

Fig. 2. RECTANGULAR.

Trenches

2' 3"
3' 9"

4' to 6' apart

Length varies with Nº of Kettles

Trench added if Time allows

2'

Kettles measure –
12 quarts 9" × 13½" × 11" high
7 do 8¼" × 12¼" × 8" do

Cooks for 360 men

Fig 3.
BROAD ARROW

8"

Direction of Wind

10

8. 45°

2.6'

Fig 4.

Raised Trench (wet weather)

+ 1.0

8

Fig 5
SECTION OF COVERED KITCHEN.

+ 6.0

3' 6

+ 1

+ 2.6

3' — 3' — 3' — 7' — 3'

Fig 6.
PLAN

3' 5

6' 6

9'

6' 6

3' 4

23

3' — 8' — 3' — 8' — 3'

4/1 8609/183 30000 7 11

Malby & Sons, Lith

Plate 35

FIELD OVENS.

Fig. I. Tracing Lines.

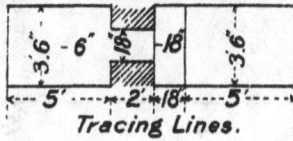

3'.6" - 6"   18"   18"   3'.6"

5'   2'.18"   5'

Tracing Lines.

Fig. 2.   Plan.

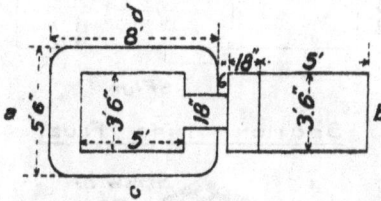

8'
d
a   5'.6"   3'.6"   18"   18"   5'   3'.6"   b
5'
c

Fig. 3. Section on a.b.

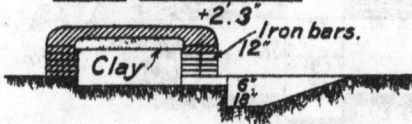

+2'.3"
Iron bars.
Clay   12"
6"
18"

Fig. 4.

Section on c.d.

+2'.3"
1'.6"

Section c.d.

OVEN BURROWED NEAR TOP OF A BANK.

Fig. 5.

1'.6"   1'9"
3'.6"

5'   Chimney optional.

Fig. 6.

Front Elevation.   Section.

Plate 36

HUTS.

PLAN OF HUT
FOR 20 MEN.

Fig. 1.

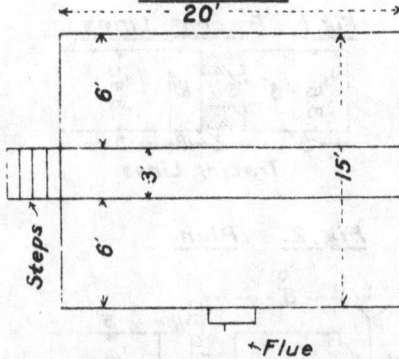

20'

6'

3'

6'

15'

Steps

←Flue

SECTION THRO FLUE.

Fig. 2.

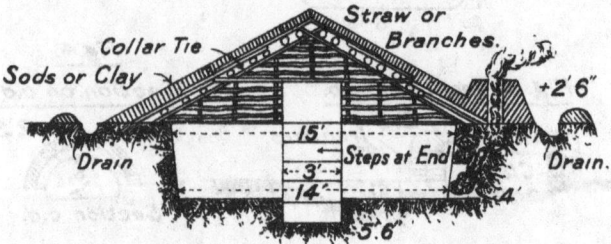

Straw or Branches.

Collar Tie

Sods or Clay

+2'6"

Drain

Drain

15'

3'

14'

4'

←Steps at End→

5.6'

Fig 3.

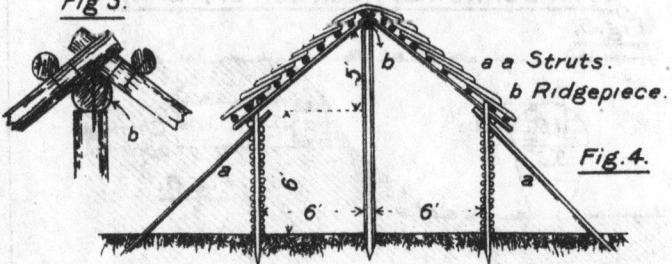

b

a a Struts.
b Ridgepiece.

Fig. 4.

b

a

a

6'

6'

6'

Plate 37

HUTS

Fig. 1
THATCHING PIECES

Fig 2
THATCHING

a a Splinters
b b Thatching Pieces

Eaves

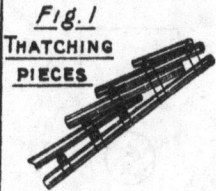

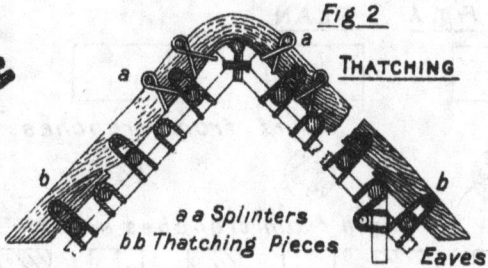

Fig 3
2 FRAMES INTERLOCKING

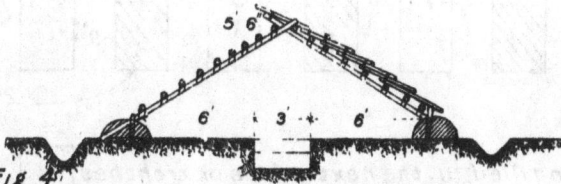

5 6
6   3   6

Fig 4
PLAN OF A FRAME

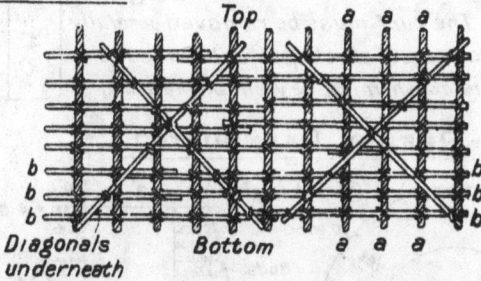

Top

a a a

b b b   b b b

Diagonals
underneath

Bottom   a a a

Fig 5.
GABLE END

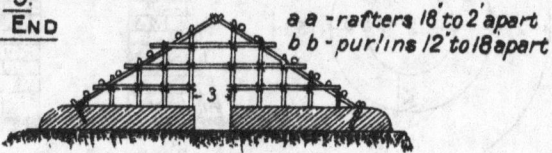

a a - rafters 18' to 2' apart
b b - purlins 12' to 18' apart

3

Plate 38.

LATRINES.

Fig. 1. **PLAN.**

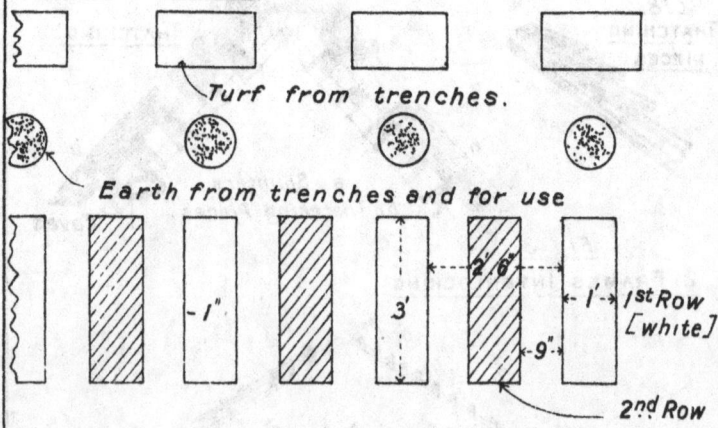

Turf from trenches.

Earth from trenches and for use

1"

3'

2'.6"

1'— 1st Row [white]

9"

2nd Row

When filled in, the next series of trenches may be made in the 2'.6" interspace if ground is limited. The turf must be removed carefully and the excavated earth put behind each trench – this earth must be well broken up.

3rd Row

REFUSE DESTRUCTOR.

Fig. 2.

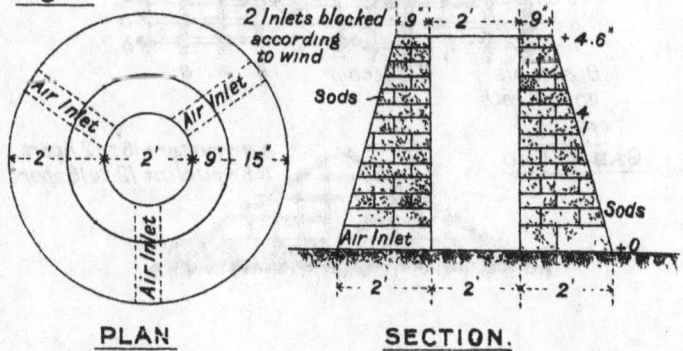

2 Inlets blocked according to wind

Sods

Air Inlet

2' — 2' — 9" — 15"

+ 4.6"

9" — 2' — 9"

Sods

Air Inlet

0

2' — 2' — 2'

PLAN **SECTION.**

Malby & Sons, Lith

Plate 39

KNOTS.

Fig 1
Thumb.

Fig 2
Figure of 8

Fig 3
Reef

Fig 4.
Single Sheet
Bend

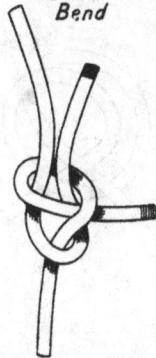

Fig 5.
Double Sheet
Bend

Fig 6.
Hawser Bend

Seizing

Fig. 7.
Commencement
of Bowline

Fig 8.
Bowline
Completed

Fig. 9
Bowline on a Bight.

471 8609/183 30000 7 11

Malby & Sons, Lith

Plate 40

KNOTS.

Fig 1
Clove Hitch

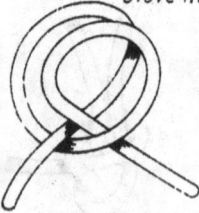

Fig 2

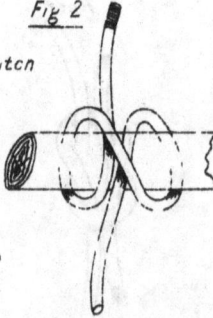

Fig 3
Timber Hitch

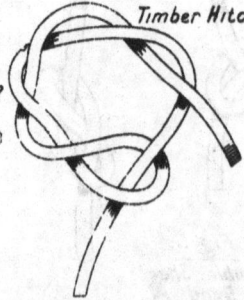

Fig 4
2 Half Hitches

Fig 5
Round Turn &
2 Half Hitches

Fig 6.
Fisherman's
Bend

Fig 8
Main Harness Hitch.

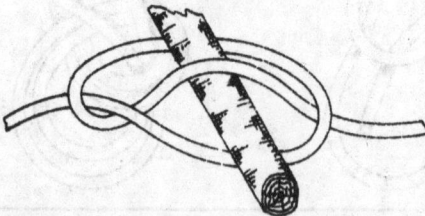

Fig 7
Lever Hitch

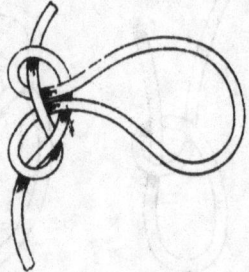

Plate 41

KNOTS

Draw Hitch.

<u>Fig. 1</u>  <u>Fig.2.</u>  <u>Fig 3.</u>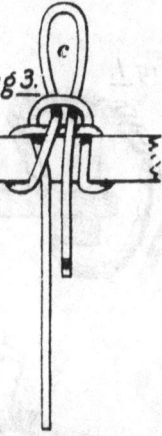

<u>Fig. 4</u> Stopper Hitch

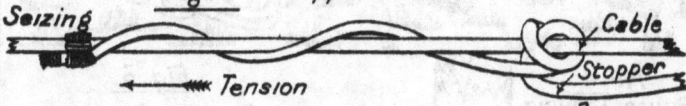

Seizing ← Tension Cable Stopper

<u>Fig. 5</u> <u>Fig. 6.</u> <u>Fig 7.</u>

Cat's Paw on Centre of Rope

Slinging a Cask Horizontally

Slinging a Cask Vertically

Plate 42.

LASHINGS.

RACK LASHINGS.

Fig 1.

Fig. 2.

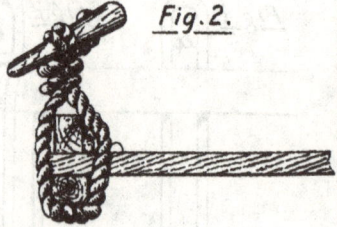

Fig. 3.
3.2.1. HOLDFAST

Guy

Fig. 4.
SQUARE LASHING
Transom

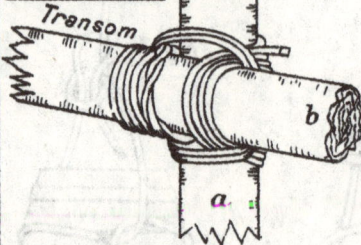

Fig. 5.
DIAGONAL LASHING

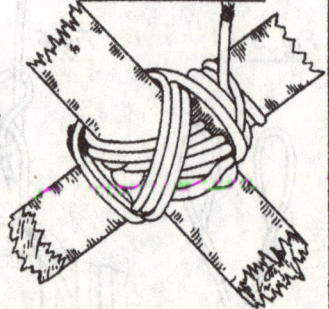

Fig. 6.
DERRICK

Fig.7. EARTH ANCHORAGE
To be filled in

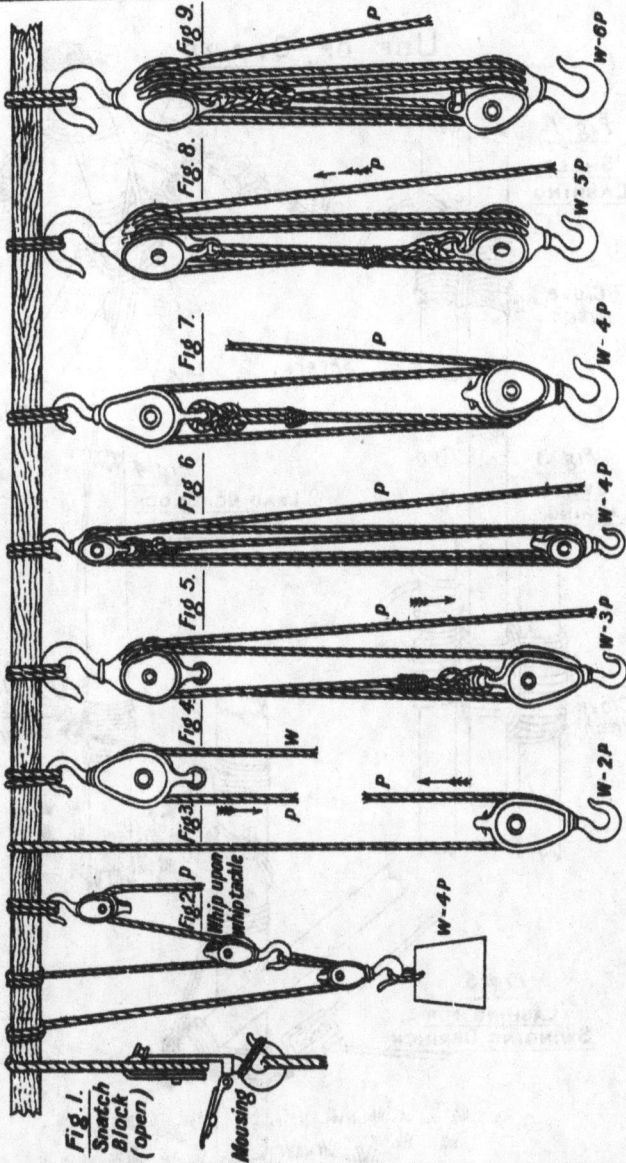

Plate 43

TACKLES.

Fig. 1. Snatch Block (open)

Mousing

Fig. 2. Whip upon Whip Tackle

Fig. 3.

Fig. 4.

Fig. 5.

Fig. 6

Fig 7.

Fig. 8.

Fig. 9.

W - 2P

W - 3P

W - 4P

W - 4P

W - 4P

W - 5P

W - 6P

P

W

471. 8808/183. 30000 7. II

Malby & Sons, Lith

Plate 44

Use of Spars

Fig 1
Sheer Lashing

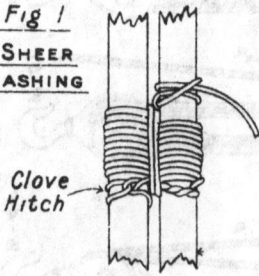

Fig. 2.

Guy

Clove Hitch

Sheers

Guy

Sling

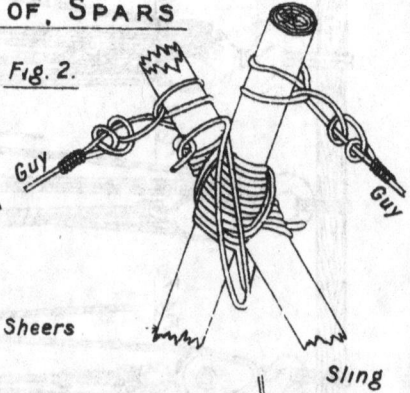

Fig 3
Gyn Lashing

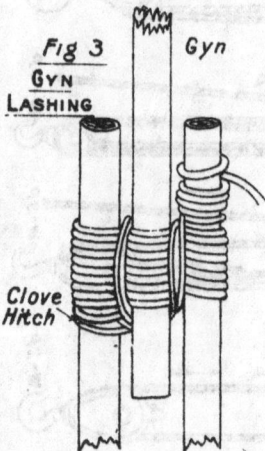

Gyn

Fig 4
Leading Block for Tackle.

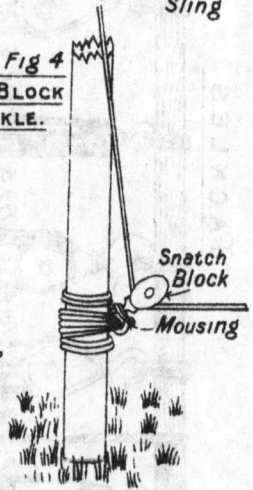

Clove Hitch

Snatch Block

Mousing

Fig 5.
Lashing for Swinging Derrick

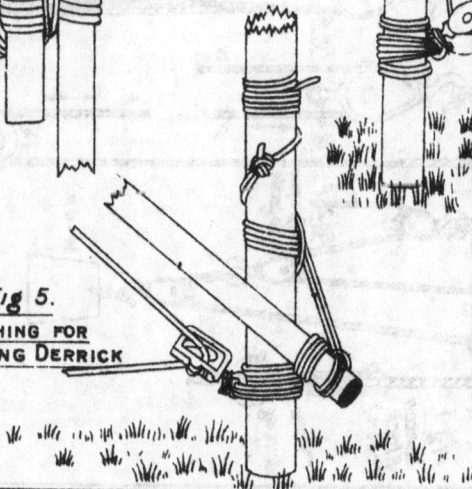

Malby & Sons Lith

Plate 45

Fig 1

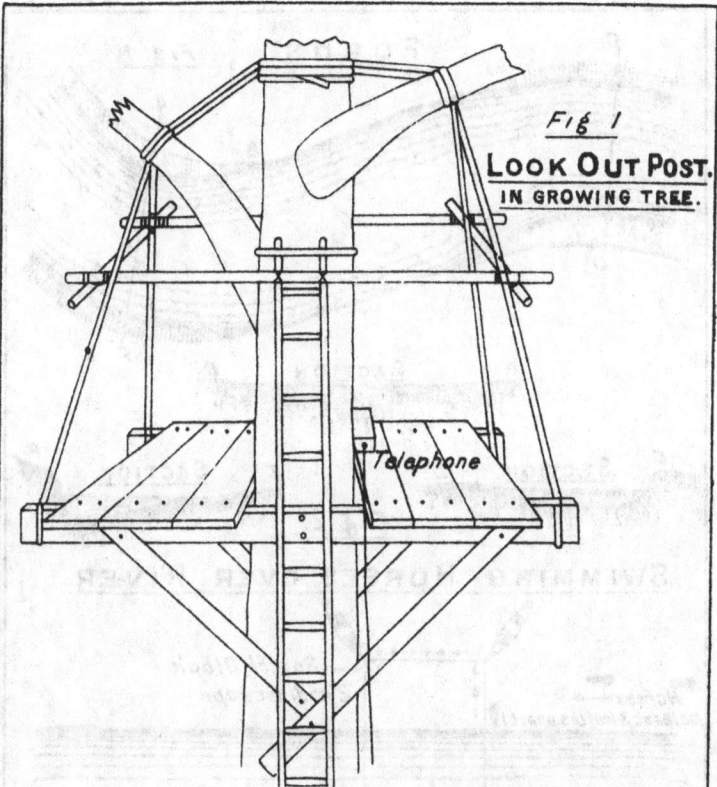

LOOK OUT POST.
IN GROWING TREE.

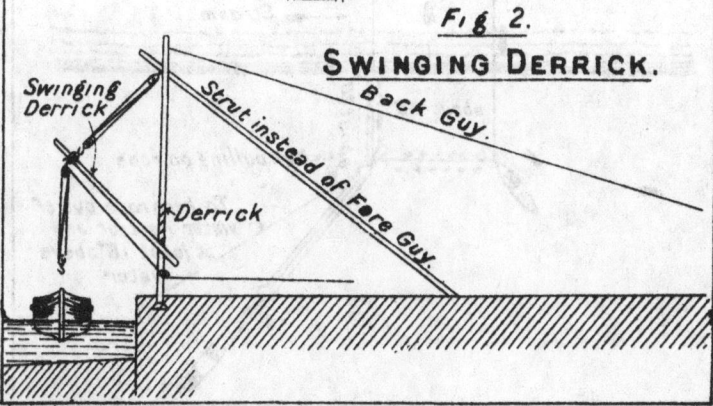

Telephone

Fig 2.

SWINGING DERRICK.

Back Guy.

Swinging Derrick

Strut instead of Fore Guy.

Derrick

471 0000/163 30000 7 11

Malby & Sons, Lith

Plate 46

FORDS. *Fig. 1.*

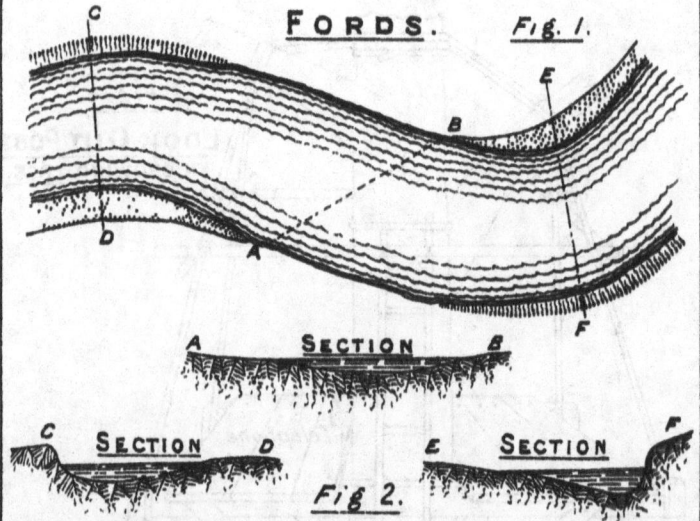

A SECTION B

C SECTION D   E SECTION F

Fig 2.

SWIMMING HORSES OVER RIVER.

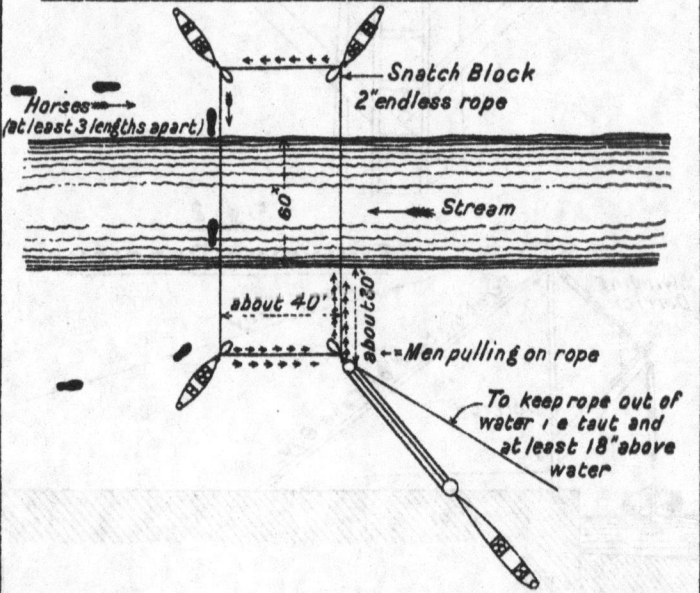

Snatch Block
2" endless rope

Horses→
(at least 3 lengths apart)

60"

←Stream

about 40'

about 30'

←Men pulling on rope

To keep rope out of water i e taut and at least 18" above water

Malby & Sons, Lith

Plate 47.

BRIDGING EXPEDIENTS.

Tarpaulin 18'×15' stuffed with straw &c

Fig. 1.

Fig. 2. Raft of four tarpaulins as Fig 1.
10 to 12 ft.

13 ft.

3'.6"

Fig. 3.

4'3"

9"

1'0"

Fig. 4.

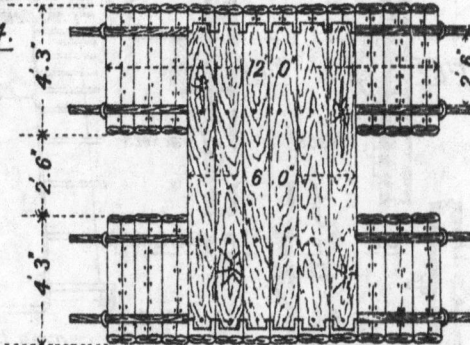

4'3"

2'6"

4'3"

2'6"

12'0"

6'0"

Raft of 24 ground sheets as Fig. 3.

Malby & Sons, Lith

Plate 48

RAPID LIGHT PILE FOOT BRIDGE.

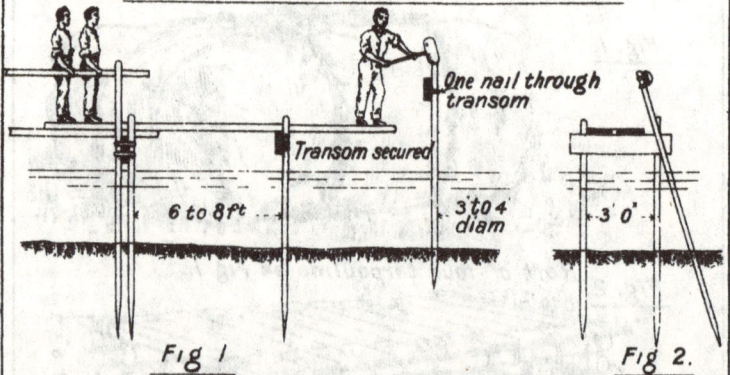

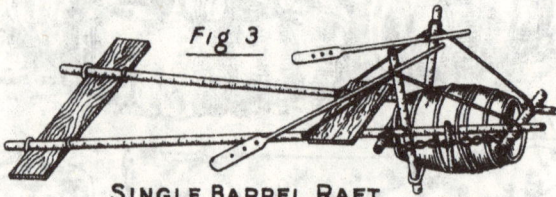

One nail through transom

Transom secured

6 to 8 ft

3' to 4" diam

Fig 1

3' 0"

Fig 2.

Fig 3

SINGLE BARREL RAFT.

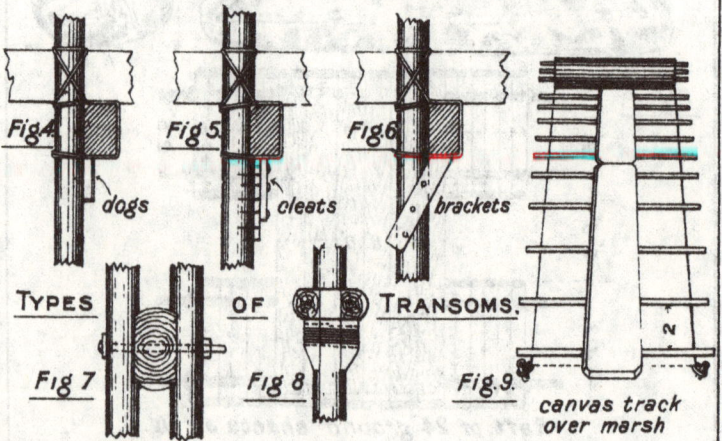

Fig 4 dogs

Fig 5 cleats

Fig 6 brackets

TYPES OF TRANSOMS.

Fig 7

Fig 8

Fig. 9.

canvas track over marsh

Plate 49

CASK FOOTBRIDGE.

PLAN.

For 54 gallon — Casks —

Fig 1

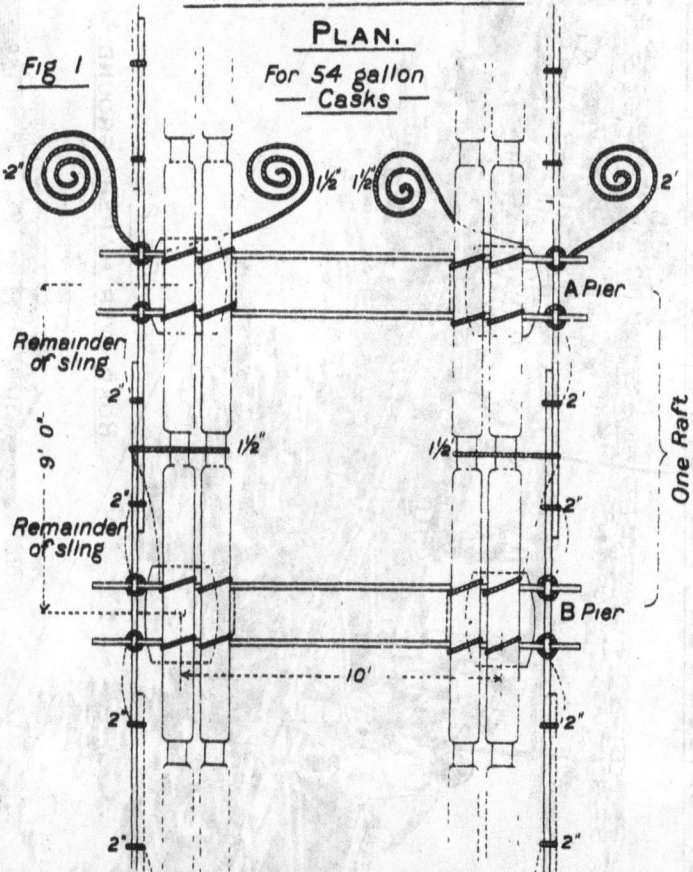

2"

1½" 1½"

2'

A Pier

Remainder of sling

9' 0"

2'

1½"

1½"

2"

Remainder of sling

2"

2'

B Pier

10'

2"

2'

2"

2"

One Raft

Where time presses, or materials are lacking planks one side only will suffice for Infantry in single file

SECTION.

Fig 2

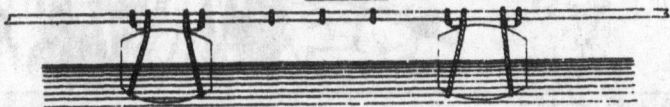

471 8609/183 30000 7 11

Malby & Sons Lith

Plate 50

CASK FOOT BRIDGE FOR PASSING ANIMALS ACROSS A RIVER.

Fig. 1.

Fig. 2

ROAD OVER MARSHY GROUND.

12'0"

2'

47/ 8609/183 30000 7 11

Malby & Sons Lith

Plate 52

TRESTLES.

Figs 1,2,3 may be either round or rectangular timber
Fig 1. And in suitable ground, the legs may be driven in and ground-
sills dispensed with

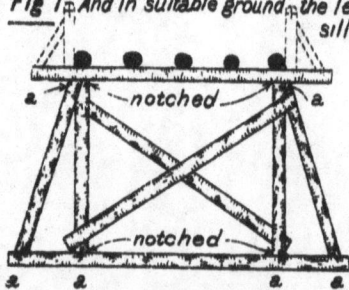

Fig. I.

"a" drift bolts

notched

notched

a a a a a a a a

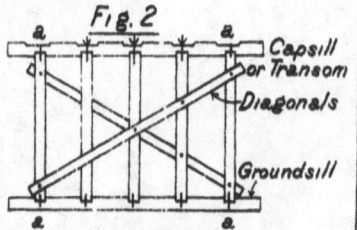

Fig. 2

a a a a

Capsill
or Transom

Diagonals

Groundsill

a a a a a a

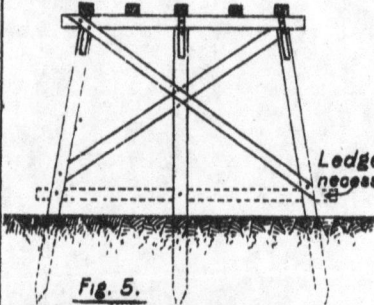

Fig. 3.

Ledger if
necessary

Fig.4
LASHED SPAR TRESTLE

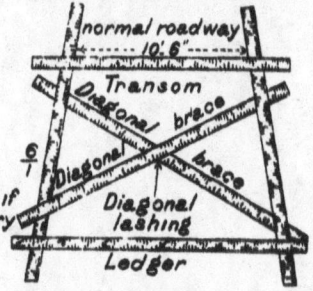

normal roadway
10'.6"

Transom

Diagonal brace

Diagonal brace

Diagonal
lashing

Ledger

All junctions lashed with wire
or rope, square lashings except
where stated.

Fig. 5.
**FOUR LEGGED TRESTLE
OF RECTANGULAR TIMBER
SPIKED TOGETHER**

Fig. 6
PLANK TRESTLE

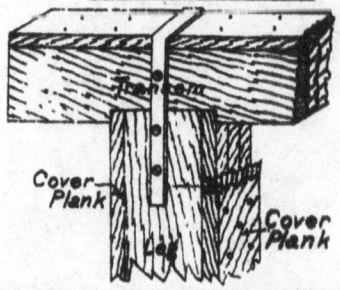

Transom

Cover
Plank

Leg

Cover
Plank

Malby & Sons, Lith

Plate 53

CANTILEVER & FRAME BRIDGES.

Fig. 1

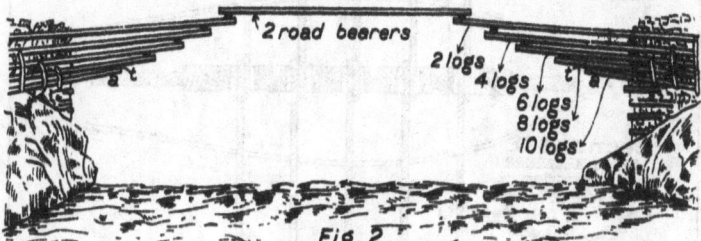

↑ 2 road bearers

2 logs
4 logs
6 logs
8 logs
10 logs

Fig. 2.

SINGLE LOCK

(No lashings shown)

a road bearers
b fork transom
c frame   "
d shore   "
e logs
f. diagonal braces

g ledger.
h footings.
i anchorage for footropes.
j chesses
k ribands.

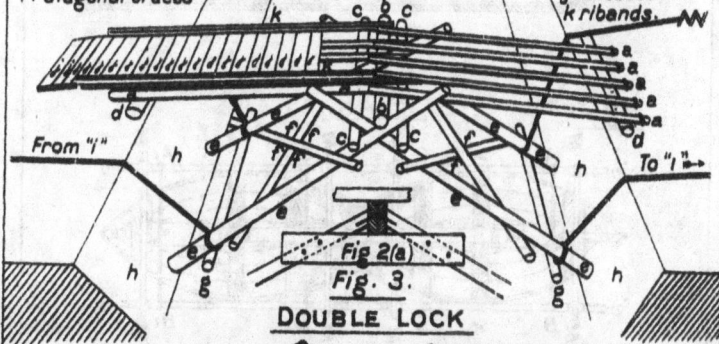

From "i"

To "i"

Fig 2(a)

Fig. 3.

DOUBLE LOCK

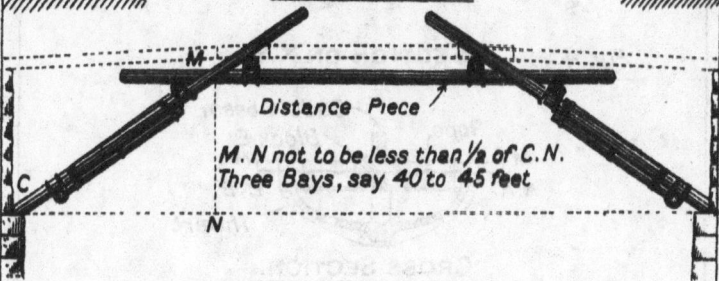

Distance Piece

M. N not to be less than ½ of C.N.
Three Bays, say 40 to 45 feet

Plate 54

BOAT PIERS.

Fig. 1. PLAN

SECTION C.D.

A
B

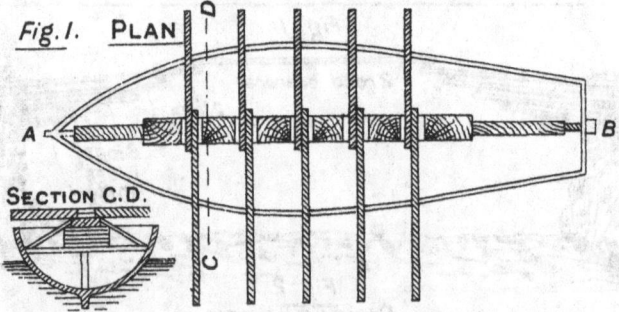

Fig. 2. SECTION ON A.B.

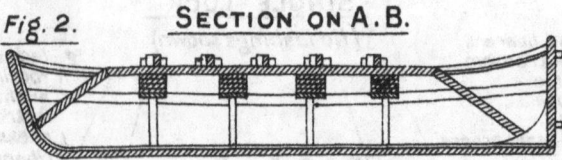

Fig. 3. PLAN.

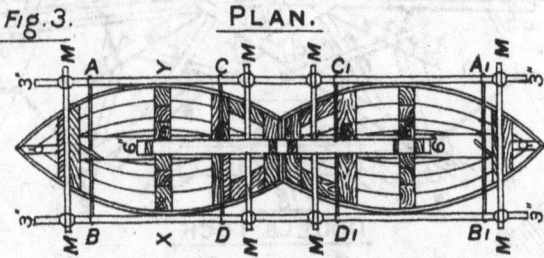

Fig. 4 SECTION ON X.Y.

Saddle beam
Rope
Block E
M
A.A
B.B
Thwart

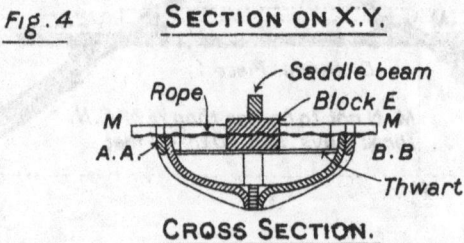

CROSS SECTION.

Malby & Sons, Lith

Plate 55

CASK PIERS.

Fig. 1

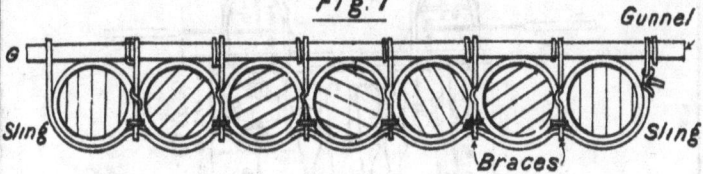

Gunnel

G

Sling

Sling

Braces

METHOD OF LASHING BARRELS TO GUNNELS.

Fig 2.

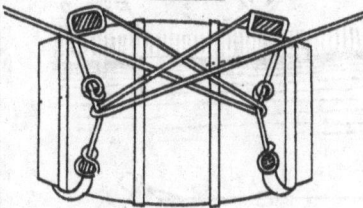

Fig. 3.

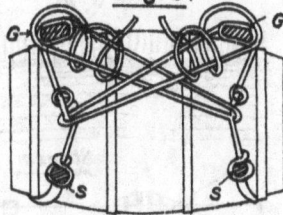

G G

S S

Fig 4.

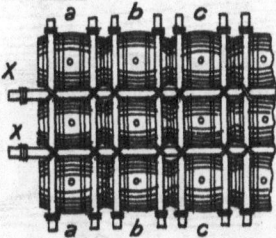

a b c

X

X

a b c

Fig. 5.

Portion of a single-pier raft which can be safely loaded shown shaded

Malby & Sons, Lith

Plate 56

ANCHORS & FLYING BRIDGES.

Fig 1.

EXAMPLE OF RAFT

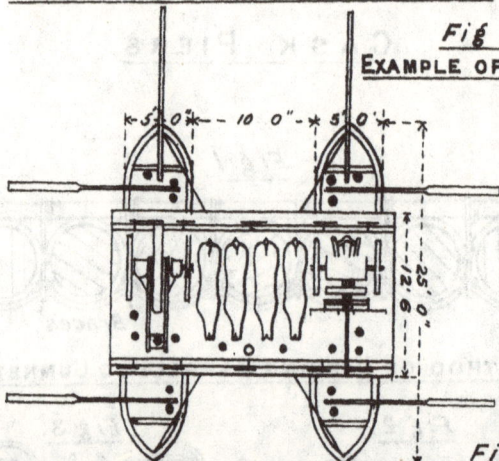

5' 0" — 10' 0" — 5' 0"

25' 0"

12' 6"

Fig. 2.

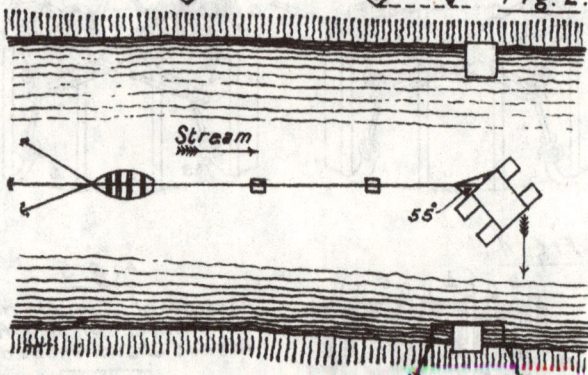

Stream

55°

Fig. 3

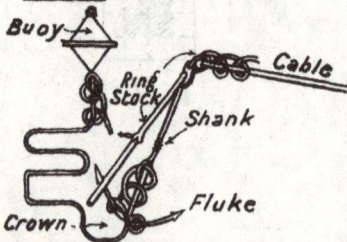

Buoy

Ring
Stock

Cable

Shank

Fluke

Crown

Fig 4

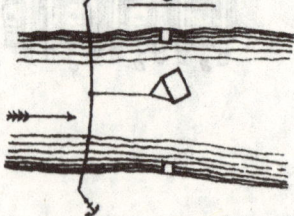

Plate 57

TABLE OF SIZES OF ROUND SPARS IN COMPRESSION, e.g. TRESTLE LEGS.

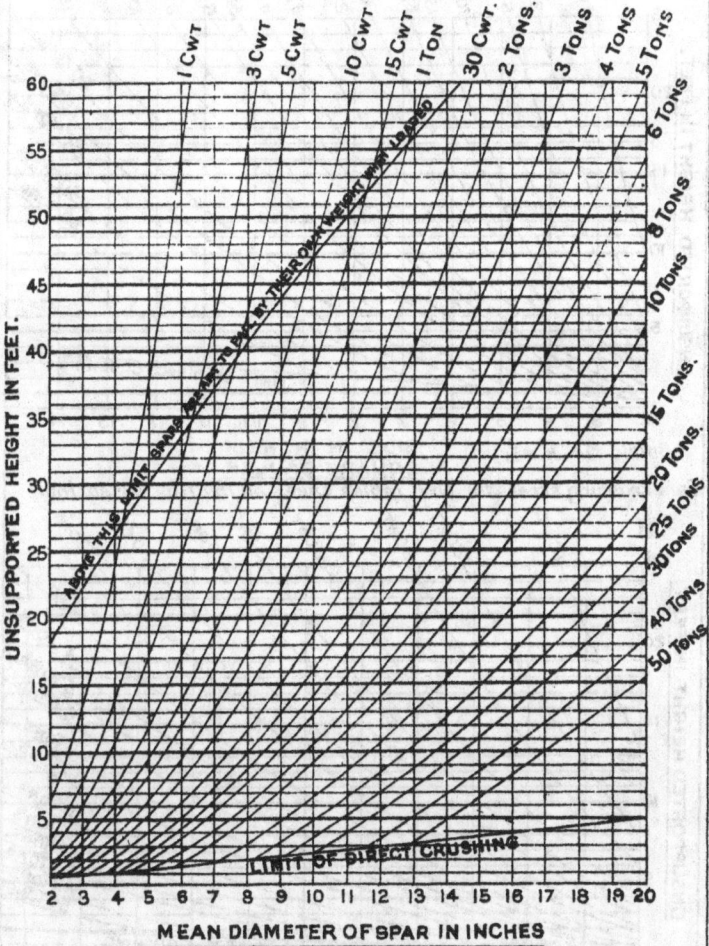

MEAN DIAMETER OF SPAR IN INCHES

Baltic fir "k"= ¼

Safe crushing stress taken as 1000 lbs. per sq. inch.

Plate 58

TABLES OF SIZES OF RECTANGULAR BAULKS
IN COMPRESSION e.g. TRESTLE LEGS.

I

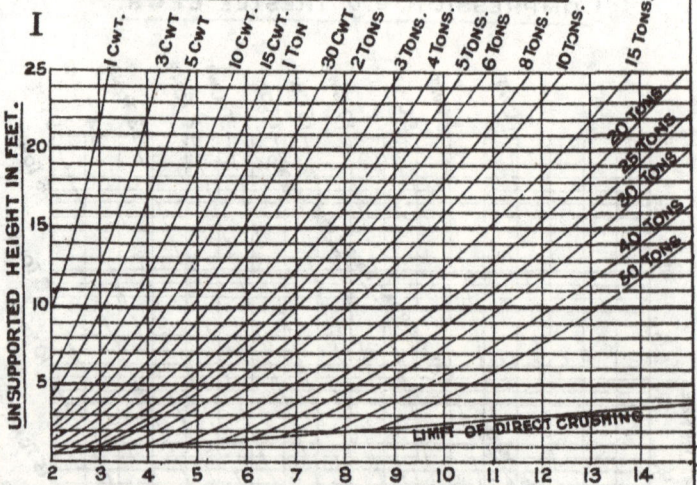

UNSUPPORTED HEIGHT IN FEET.

1 CWT. · 3 CWT · 5 CWT · 10 CWT. · 15 CWT. · 1 TON · 30 CWT · 2 TONS · 3 TONS. · 4 TONS. · 5 TONS. · 6 TONS. · 8 TONS. · 10 TONS. · 15 TONS.

20 TONS. · 25 TONS · 30 TONS. · 40 TONS. · 50 TONS.

LIMIT OF DIRECT CRUSHING

Baltic fir "k" = 5/4

SIDE IN INCHES
SQUARE BAULKS

Safe crushing stress for both Tables taken as 1500 lbs. per sq. inch

II.

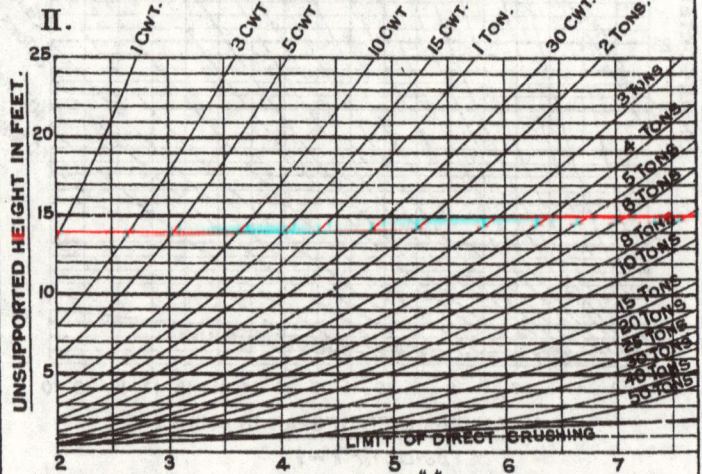

UNSUPPORTED HEIGHT IN FEET.

1 CWT. · 3 CWT · 5 CWT · 10 CWT · 15 CWT. · 1 TON. · 30 CWT · 2 TONS.

3 TONS · 4 TONS · 5 TONS · 6 TONS · 8 TONS · 10 TONS · 15 TONS · 20 TONS · 25 TONS · 40 TONS · 50 TONS

LIMIT OF DIRECT CRUSHING

Baltic fir "k" = 5/4

LESSER SIDE (b) IN INCHES.
RECTANGULAR BAULKS SIDES 2:1 (a = 2 b)

47. 8603/183 30,000 7/11

Malby & Sons, Lith

Plate 89

HASTY DEMOLITIONS.

SECTION OF
(Nº 8 Mark IV) DETONATOR

Fig.1.

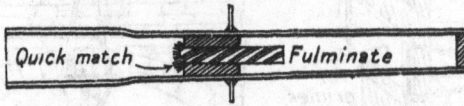

Quick match ————— Fulminate

METHOD OF JOINING FUZES

Fig.2. Scarf Joint

Safety Fuze   Instantaneous Fuze

Fig 3.

Semi-circular nick
in Safety Fuze.

Fig. 4.

Semi-circular nick in
Instantaneous Fuze.

Fig 5.

Nicked joint between Safety & Instantⁱ Fuze.

SECTION OF
Fig.6. COMMERCIAL DETONATOR

Fulminate

Fig.7.

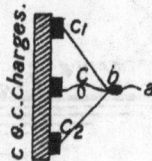

c.c.c.charges.

C₁

C

b   a

C₂

471 8608/183 30000 7 11

Malby & Sons, Lith

Plate 60

HASTY DEMOLITIONS.

See Section 109

Fig. 1.

Fig 2

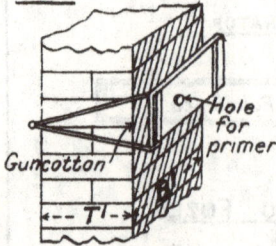

Guncotton — Hole for primer

Charge $\frac{2}{3}$ BT² lbs

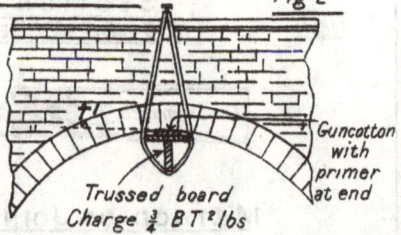

Trussed board
Charge $\frac{1}{4}$ BT² lbs

Guncotton with primer at end

Fig. 3.

Fig. 4.

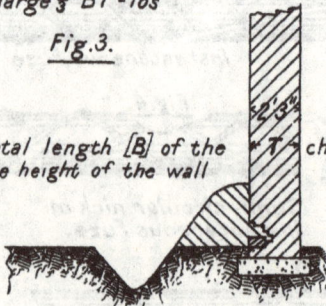

Total length [B] of the the height of the wall

Charge $\frac{1}{4}$ BT² lbs (untamped.)

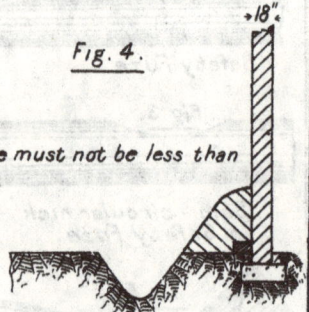

→18″

"T" charge must not be less than

Charge 2 lbs per ft. run (untamped.)

Half these amounts of Guncotton if tamped.

Fig. 5.

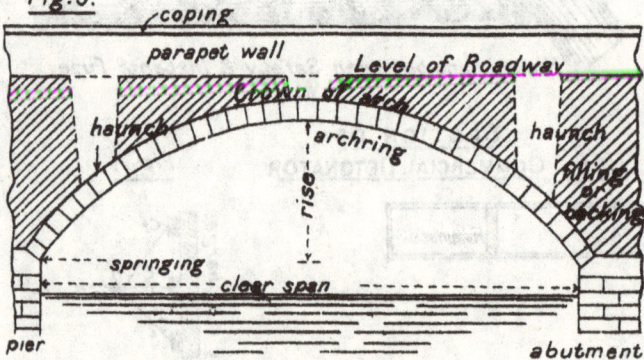

coping
parapet wall
Level of Roadway
Crown of arch
haunch
archring
haunch
filling and backing
rise
springing
clear span
pier
abutment

Plate 61

PREPARING CHARGES.

Fig 1 — Fuze — Plank — Guncotton — Stockade

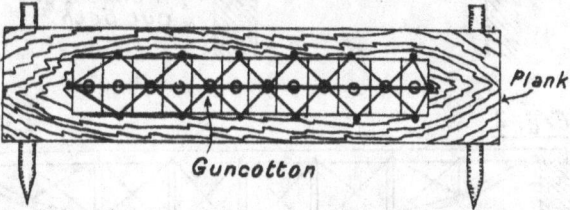

Fig. 2. — Plank — Guncotton

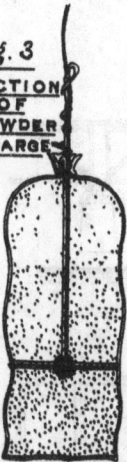

Fig. 3
SECTION OF POWDER CHARGE

A service sandbag holds about 40 lbs. of Gunpowder.

Fig. 4.
SAND BAG FILLED

Fig. 5.

Sandbag — Sandbag — Sandbag — Sandbag — Sandbag

Direction of wind

Fuze inserted into powder bag

Malby & Sons, Lith.

Plate 62

GIRDER BRIDGES

Fig. 1.

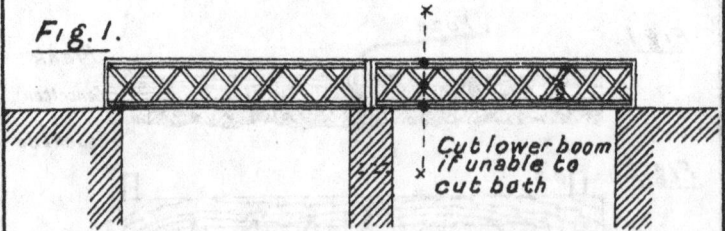

Cut lower boom
if unable to
cut both

Fig. 2.

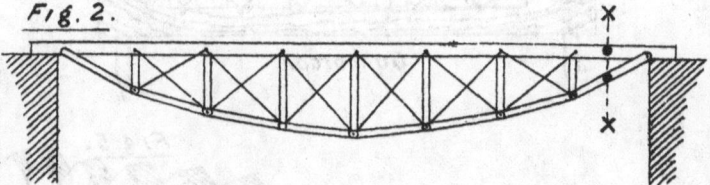

Fig. 3.

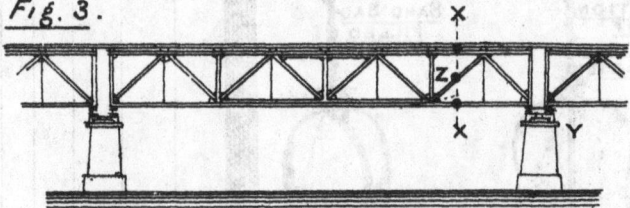

Fig. 4.

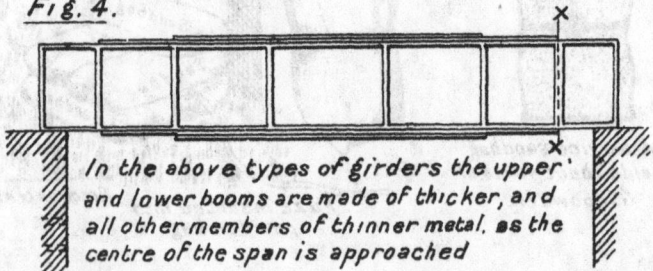

*In the above types of girders the upper
and lower booms are made of thicker, and
all other members of thinner metal, as the
centre of the span is approached*

471 8609/183 30000 7 11   Malby & Sons. Lith

Plate 63

DEMOLITION OF RAILWAYS.

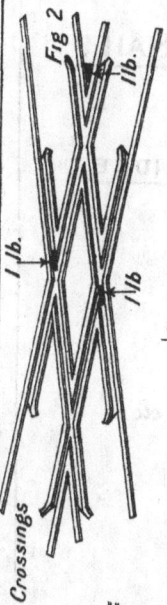

Fig.1.

Points 1lb.

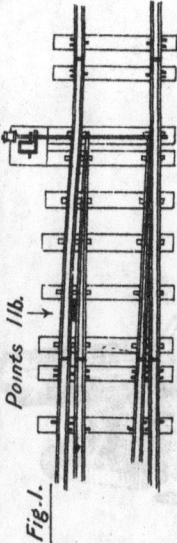

Fig 2

Crossings

1 lb. 1lb. 1lb. 1lb.

MATERIALS :–
String 3ft.
Clay
Stick
Weight
Matches
Knife

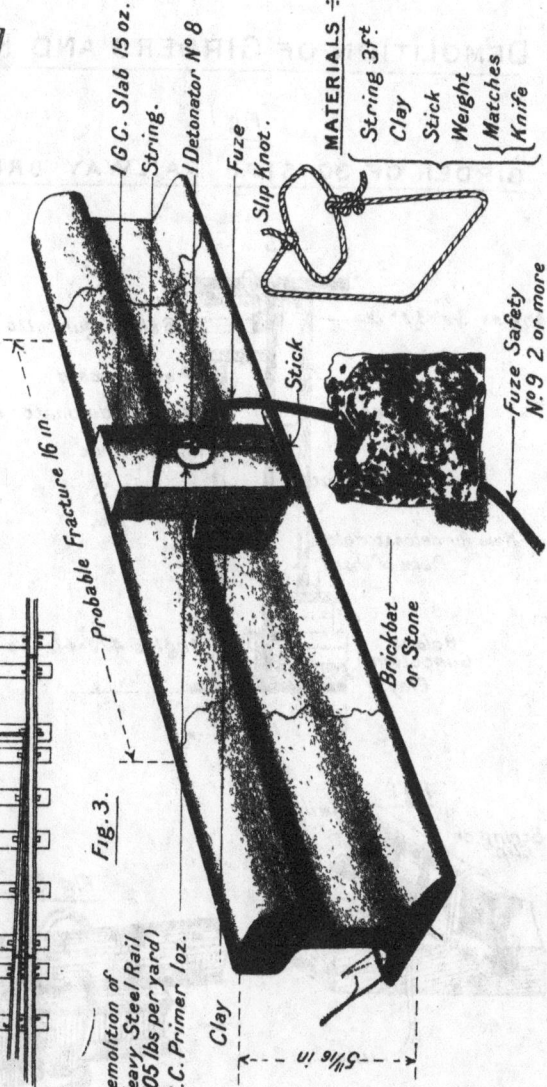

1 G.C. Slab 15 oz.
String.
1 Detonator Nº 8
Fuze
Slip Knot.
Stick
Fuze Safety
Nº 9 2 or more
Brickbat or Stone
Clay
5·9/16 in.
Probable Fracture 16 in.
Fig.3.
Demolition of Heavy Steel Rail (105 lbs per yard) 1 G.C. Primer 1oz.

471 8603/183 30000 7 11

Malby & Sons, Lith

Plate 64

DEMOLITION OF GIRDERS AND RAILS.

Fig 1

GIRDER OF 30´ SPAN RAILWAY BRIDGE.

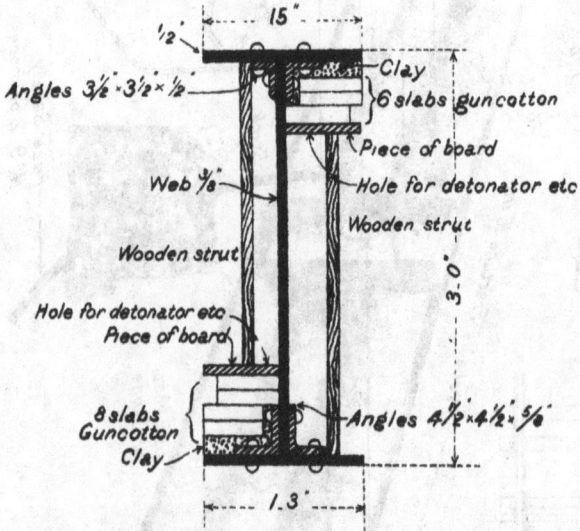

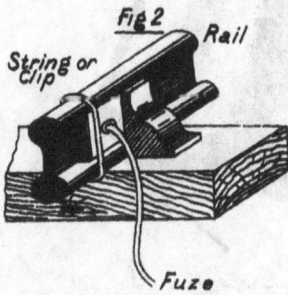

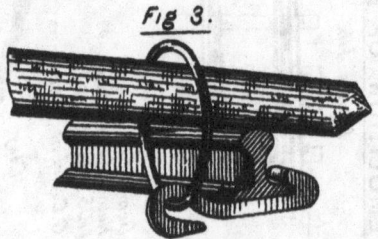

15"
½"
Angles 3½"·3½"×½"
Clay
6 slabs guncotton
Piece of board
Web ⅜"
Hole for detonator etc
Wooden strut
Wooden strut
Hole for detonator etc
Piece of board
3·0"
8 slabs Guncotton
Clay
Angles 4½"×4½"× ⅝"
1·3"

Fig 2
Rail
String or Clip
Fuze

Fig 3.